William D. Roebuck, William E. Clarke

A Handbook of the Vertebrate Fauna of Yorkshire

being a catalogue of British mammals, birds, reptiles, amphibians, and fishes

William D. Roebuck, William E. Clarke

A Handbook of the Vertebrate Fauna of Yorkshire
being a catalogue of British mammals, birds, reptiles, amphibians, and fishes

ISBN/EAN: 9783337272791

Printed in Europe, USA, Canada, Australia, Japan

Cover: Foto ©berggeist007 / pixelio.de

More available books at **www.hansebooks.com**

A HANDBOOK

OF THE

VERTEBRATE FAUNA

OF

YORKSHIRE

BEING A CATALOGUE OF BRITISH MAMMALS, BIRDS, REPTILES, AMPHIBIANS,
AND FISHES, SHOWING WHAT SPECIES ARE OR HAVE, WITHIN
HISTORICAL PERIODS, BEEN FOUND IN THE COUNTY.

BY

WM. EAGLE CLARKE,

MEMBER OF THE BRITISH ORNITHOLOGISTS' UNION,

AND

WM. DENISON ROEBUCK;

THE SECRETARIES OF THE YORKSHIRE NATURALISTS' UNION.

LONDON :
LOVELL REEVE & CO., HENRIETTA ST., COVENT GARDEN.
LEEDS: RICHARD JACKSON, COMMERCIAL STREET.
1881.

TO

Sir JOHN LUBBOCK, Bart., M.P.,

D.C.L., LL.D., F.R.S., &c.,

AS

President of the British Association,

ON THE OCCASION OF

Its meeting in the City and County of its Origin,

TO CELEBRATE THE COMPLETION OF

The First Fifty Years of its Existence,

This contribution to the Natural History of the County

IS DEDICATED.

INTRODUCTION.

THE object of the present work is the enumeration of the vertebrated animals which are or have been found in Yorkshire, and the careful definition of their faunistic position and geographical distribution within the county, in language as terse and as accurate as it is possible to employ. Such a work has never been undertaken for the county, nor indeed has there been published a list either of the vertebrata as a whole, or of any of the classes into which the sub-kingdom is divided. In this respect Yorkshire affords a marked contrast with the neighbouring counties of Norfolk and of Northumberland and Durham, whose avifaunas especially have been written—and more than once—by competent and able ornithologists.

The number of British vertebrata which have not occurred in Yorkshire being comparatively small, it seemed desirable to make the work not only a county handbook, but a complete nominal catalogue of the British species. Such a catalogue is in itself a desideratum, especially if brought up to the standard of present knowledge, and will be of use both as furnishing a ready means of comparison and as facilitating the registering of additions to the Yorkshire fauna. Careful attention has been paid to the classification and nomenclature, both being based upon the works of the most recent and reliable authorities.

Mammalia.—The list of British mammalia here given is substantially that of the second edition of Bell's History of British Quadrupeds (1874), with such slight modifications as to classification and nomenclature as are necessary to bring it down to the present time. The list of Bats is entirely based upon Dr. G. E. Dobson's British Museum Catalogue of Chiroptera, published in 1878, and the classification of Seals and Cetaceans is that of Prof. Flower, as given in Bell's work, and in Mr. Southwell's recently published 'Seals and Whales of the British Seas'; while papers by Mr. St. George Mivart on the Insectivora, Mr. E. R. Alston on the Rodentia, and Prof. Flower on the Carnivora, have been consulted as to the arrangement of those respective groups. Two species given by Bell are omitted from the present list, and one is added. The latter, a bat—*Vespertilio dasycneme*—is included on the authority of Dr. Dobson's monograph. The species omitted are the Beech Marten and the Greenland Right-Whale. Mr. E. R. Alston has demonstrated that there is but one British species of Marten, the true Beech Marten never having occurred (P.Z.S., 1879, p. 468 ; Zool., 1879, p. 441). The Greenland Right-Whale (*Balæna mysticetus* L.) has hitherto been included in the British fauna upon evidence so unsatisfactory that modern faunists invariably express grave doubts as to the validity of its claims. It has therefore been deemed the wiser plan to omit it altogether, the strong probability being that all Right-Whales killed in British seas have been referable to *B. biscayensis*, a species whose differentiation as such dates subsequently to all the records of British occurrences.

As to the extinct British mammalia, it having been considered desirable to include notices of animals which had ceased to exist in Yorkshire within historical periods, the species—five in number —of which Mr. J. E. Harting treats in his work on 'Extinct British Animals,' have been inserted in the list in their correct zoological sequence. Their names, and that of the Great Auk, are, however, printed in old English characters, and left unnumbered, as not being now entitled to rank as true members of the British fauna.

Birds. — When considering what system of classification should be adopted for the arrangement of this class, difficulty was experienced in coming to a decision, and it was only after some hesitation that one was finally arrived at. Had Professor

Newton's admirable edition of Yarrell's 'British Birds' been complete—or nearly so—it would undoubtedly have been adopted as a guide by virtue of the position which it admittedly holds as the standard work on British Ornithology. In such an event practical convenience would—and rightly so—have overruled all other considerations. But this unfortunately not being the case, it became necessary to consider the present state of ornithological opinion as to the classification of Birds. This has been admirably summarised in the Ibis for 1880, by Dr. P. L. Sclater, in a paper entitled 'Remarks on the present state of the Systema Avium.' The arrangement therein promulgated—or some modification of it—seems likely to meet with acceptance more or less general at the hands of ornithologists, and has already been adopted by Mr. H. E. Dresser in his recently-completed great work on the Birds of Europe. The arrangement and nomenclature of Mr. Dresser's work have therefore been followed.

It is not within the province of a work of this character to decide whether certain birds ought or ought not to be included in the list, and consequently all admitted as British by Mr. Dresser, or included by Mr. Wharton in his excellent little catalogue, have, with trifling exceptions, been here included. There are, however, certain species which have been reported as occurring in Yorkshire on evidence which is regarded by authors as more or less insufficient for their admission into the British fauna. These are inserted in their zoological position, but—in order that it may be quite clear that their claims are not fully accepted —no numbers are prefixed. Nor are they prefixed to *Cygnus olor* and *Alca impennis*—the first a domesticated bird and the second admittedly extinct.

Reptiles and Amphibians.—The lists of reptiles and amphibians are founded upon Bell's British Reptiles and the writings of Dr. Günther and Mr. St. George Mivart. An amphibian, however, which has been described in the works of Bell and Cooke—*Ommatotriton vittatus* or Gray's Banded Newt, a Syrian form—is here omitted, as it has been shown to have no claim to a place in the British list, into which it had been introduced on the strength of museum specimens.

Fishes. — It is fortunate—so far as the classification and nomenclature are concerned—that the list of British Fishes can be based upon reliable and satisfactory authority. The classification adopted is that promulgated by Dr. Günther in his recently published 'Introduction to the Study of Fishes'; while his well-known 'Catalogue of Fishes' furnishes a safe guide to the nomenclature. Dr. Day's papers in the Linnean Society's Journal, and more particularly his comprehensive work on the 'Fishes of Great Britain and Ireland,' have also served as a guide to some of the conclusions arrived at.

The British Fishes here enumerated are substantially those of the third edition of Yarrell's 'History of British Fishes' (1859), modified by the assistance of the writings of the two distinguished ichthyologists just referred to. Some few, however, of the species included by Yarrell, and a considerable number of those added by Couch, have—since the date of their writings—proved to be either monstrosities, varietal or immature forms of other species, or to have been admitted into the British fauna on insufficient evidence. Their omission consequently requires no further explanation.

In attempting to define the faunistic position of fish, not only is there found a deficiency of the requisite information for the Yorkshire coast, but considerable dissatisfaction is the usual result of a reference to the works of Yarrell and Couch. Admirable as they are in certain respects, the vagueness of some of the statements made, and the want of system in the arrangement of the subject-matter, often renders it extremely difficult to ascertain the geographical range of a species, still more so to learn its true position in the British fauna. This remark applies even in the case of some of the commonest forms. It is true that the habits of fishes are very obscure and their natural history but imperfectly ascertained, but in the case of common species this unsatisfactory nature of the literature is unjustifiable. It is, however, a matter of sincere gratification to find that the important want complained of is to a great extent supplied, and that a systematic arrangement of the subject-matter is adopted in Dr. Day's new work, one which is destined to be of great service to the British Ichthyologist.

The Faunistic Position of Species.—The most important requirements in the compilation of a local fauna are a careful definition of the true faunistic position occupied by each species, and of its distribution and relative numbers within the area treated of, together with some notice of its migratory movements. To these should be added—in the case of the rare species—lists of all the occurrences, with dates, localities, authorities, and such other details as are likely to be of service.

For the purpose of defining the faunistic position of the Yorkshire Vertebrata the following terms are employed :—

RESIDENTS . . . ⎫
SUMMER VISITANTS . ⎬ = ANNUAL BREEDERS.
⎧

WINTER VISITANTS . ⎫ = REGULAR VISITORS.
PERIODICAL VISITANTS . ⎭

CASUAL VISITANTS . . ⎫ = IRREGULAR VISITORS.
ACCIDENTAL VISITANTS ⎭

These terms are more easily applicable to the birds than to other classes of the vertebrata, from the greater facility with which their movements can be observed and noted. In the case of the marine fishes and cetaceans the terms would be quite as appropriate, were it not for the veil of obscurity which invests their habits, and in a lesser degree the small attention which they receive at the hands of local naturalists. Owing to this consideration, it is advisable in defining—and more especially in illustrating —the terms, to speak of them as applied to birds.

RESIDENTS are species which are found in some district or other of the county throughout the year, and therein breed annually.

SUMMER VISITANTS are species which appear annually in the spring, remain through the summer for the purpose of rearing their young, and afterwards depart in the autumn.

WINTER VISITANTS are species which appear annually in the autumn, and remain in more or less numbers throughout the winter, departing in the spring for their breeding haunts.

PERIODICAL VISITANTS are species which are observed in the county only on their annual passage to and from their breeding haunts in spring or autumn or both.

CASUAL VISITANTS are species whose appearance in the county is uncertain, but whose occurrence—they being resident in, or more or less regular visitants to, other parts of the British Isles —is not improbable, even though their visit may be very few and far between.

ACCIDENTAL VISITANTS are mere waifs and strays—species whose geographical range renders their occurrence in Britain quite exceptional and more or less remarkable.

These definitions have been carefully framed, and will, it is believed, be found applicable to all cases. A few general remarks upon them, illustrated by characteristic examples, desirable in order to make their meaning perfectly unmistakable, will be given in proceeding to analyse the Yorkshire fauna.

YORKSHIRE:

ITS PHYSICAL ASPECT AND VERTEBRATE
FAUNA.

YORKSHIRE:

Its PHYSICAL ASPECT and VERTEBRATE FAUNA.

YORKSHIRE, the largest county of the British Isles, containing an area of 3,936,242 statute acres, or 6150 square miles, and situate between 53° 18' and 54° 40' N. latitude and about 9' E. and 2° 36' W. longitude of the meridian of Greenwich, is also one of the most compact in form, the most varied in geological structure, soil, climate, and physical aspect.

The lands of Yorkshire rise in masses from S.E. to N.W., in a direction which corresponds with that of the age of the underlying rocks, the oldest or palæozoic formations constituting the high mountains of the north-west, whilst the newest or tertiary deposits of Holderness occupy the opposite or south-east angle. Thus a line drawn from the beach at Spurn to the highest summit of Yorkshire—Mickle Fell, 2596 feet—marks not only the general slope of the high lands but their succession in geological time, and is moreover the longest line (120 miles) that it is possible to draw within the county.

Broadly speaking the most salient features of its physical configuration are the great central depression and the flanking masses of hills to the east and west.

The North-Western Fells is a wild and picturesque tract of mountainous country, ascending to 2596 feet at the extreme north-western angle of the county, and nowhere descending to a lower elevation than about four hundred feet. A district of lofty hills, thirty-six of which attain an altitude of two thousand feet or more, of extensive stretches of heathery moorlands, of grassy slopes and grey limestone scars, diversified by waterfalls, caves, clear and sparkling streams, and beautiful and romantic dales, this elevated region includes the main watershed of the North of England, and within its limits and upon Yorkshire soil rise all the great rivers of the north—Tyne and Wear alone excepted.   The steep western slopes are drained into the Irish Sea by the Eden, the Lune, the Ribble, and their tributary streams ; while down the broader valleys and more gentle inclines of the eastern slopes flow the Aire, the Wharfe, the Nidd, the Ure, the Swale, and the Tees, into the North Sea.

The gritstone summits and limestone scars of this region are the last refuge in Yorkshire of the buzzard, and amongst the last of the raven and the peregrine ; the high moors are inhabited by the red grouse, ring ouzel, merlin, twite, curlew, dunlin, snipe and golden plover, while the dipper, grey wagtail, and sandpiper are abundant on the mountain becks.   The rivers and streams of the district are inhabited by little else than trout, and such salmon and sea-trout as are able to pass the dams and weirs which for the most part bar their ascent of the Yorkshire streams.

Of the very few natural sheets of water in Yorkshire this district possesses three of the most important, Malham Tarn, Semer-water, and Birkdale Tarn, besides a few others of smaller size. Malham Tarn, 153 acres in extent and 1,250 feet above the level of the sea, together with the limestone plateau on which it is situate, is of special interest as illustrating the altitude to which certain species will ascend, and its fauna has therefore been made a special feature in the list: Here the wood-wren, redshank, teal, coot, and dabchick nest annually, and it is the only locality in Yorkshire where the tufted duck has been known to breed ; while ichthyologically its fauna is remarkable for the malformation of the trout, in all probability the result of isolation.

Of the mammalian fauna which formerly inhabited this wild and secluded mountain district but little is known.   The dense woods of the ' forests ' of Upper Teesdale, Richmondshire, and Wensleydale, are known to have been the haunt of the wolf, red deer, and other beasts of chase, of whose extinction little is known, the actual records of their former presence being very scanty.   The wild cat, bear, and wild boar no doubt also existed there, but of them no record remains.   The marten possibly still survives, as within quite recent years examples have been killed at Buckden and Azerley; and probably the last remnant of the ancient fauna is to be found in the small herd of red deer which are still preserved in the Deer Park at Bolton—doubtless the lineal descendants of those which roamed in vast herds over the whole district in days gone by.

The Craven Pasture-lands.—Immediately below the North Western Fells, which are abruptly terminated to the south by the steep and occasionally precipitous descents of the Craven and Pennine faults, succeeds a comparatively low region, under 600 feet in elevation, with an undulating grassy surface and low rounded hills, in places rising into fells which reproduce on a smaller scale the leading physical characteristics of those of the north-west.   Through the green pastures of this uninteresting country, of which the peewit is the characteristic bird, the Ribble and the Hodder cut their way in the form of narrow, well-wooded, sheltered, and productive ravines, which give some charm to this otherwise monotonous country.

Formerly constituting the famous forest of Bowland, this district is chiefly of interest as the last part of the county in which the wild white oxen survived.   A herd was for a long time perpetuated at Gisburn Park, but the last was killed in 1859, on account of the degeneration of the race, resulting from many centuries of interbreeding and isolation.   In this district also are the only localities for the natterjack toad, which here occurs sparingly, and for the whiskered bat, of which a single specimen was taken—the only one known for Yorkshire.

The South-Western Moorlands.—The summit ridge, broken and irregular among the fells of the north-west, and interrupted by comparatively low ground south of them, begins again near Keighley and Ilkley, and is carried southward by a broad and continuous band of elevated and monotonous rolling heatherland, which extends along the county boundary as far as Derbyshire, and attains its greatest elevation—1859 feet—at Holme Moss.   These unbroken stretches of dreary moorlands — unrelieved save by deep and narrow 'cloughs' or ravines, are, in comparison with the Fells of the north-west, of but slight interest to the naturalist. Homogeneous in their geological structure, and presenting no other soils than the barren and unproductive peat-laden and heather-covered millstone grit, they afford little variety in their fauna. The high moors are inhabited by grouse—more strictly preserved here than elsewhere—and by occasional pairs of curlew, golden plover, snipe, black grouse, ring ouzel, and less frequent still an odd pair of dunlin; the streams are the haunt of the dipper, the grey wagtail, and sandpiper, while the lower parts of the valleys are inhabited by such birds and animals as are able to maintain their ground against man and his works. For the south-western moorlands are situate between the two great coalfields and manufacturing districts of Yorkshire and Lancashire, and are not only of easy access to a vast population, but within the direct influence of the clouds of smoke which accompany the manufacture of cotton upon the one side, and woollens and worsteds upon the other.

The Manufacturing District.—At the foot of the south-western moorlands, and to the east of them, the great Yorkshire coalfield stretches from Leeds and Bradford to Halifax, Huddersfield, Wakefield, Barnsley, and Sheffield. Within this comparatively limited area is congregated the great mass of the population of Yorkshire, for here the presence of coal and ironstone has determined the location of some of the world's greatest industries; and the coal-mining districts of the West Riding afford one of the clearest demonstrations of the transforming influence of human agencies upon the surface of a country. The air is laden with smoke above, the rivers run black and polluted below, vegetation is checked and stunted, animal life is scarcely able to maintain its ground, and fish have long been banished from rivers whose foulness and inky blackness can only be paralleled by that of the streams of the neighbouring county palatine of Lancaster.

Naturally well-wooded, the district still retains that characteristic in parts, more especially in the southern portion, where the noble Chase of Wharncliffe, overlooking an extensive prospect in the Don valley, and the fine parks about Barnsley and Wake-field, still afford a shelter to woodland species of birds, some of considerable interest—such as the nightingale and the pied fly-catcher—though the inimical influence of smoke has long told upon the trees. The district is also interesting as within it is situated Walton Park—the sylvan domain wherein for many years Charles Waterton extended complete protection to living things of all kinds. Here flourished a famous heronry, which after the death of Mr. Waterton was disturbed, and finally dispersed. But the continued presence of so dense a population and the ever-increasing demands of modern commerce are gradually breaking up and destroying what suitable habitats the district still possesses, hastening the process of extinction which is continually going on, and thus diminishing a fauna which was never at any time a very rich one.

The Central Plain, including under this name not only the entire vale of York, but also the lowlands of Cleveland and the Tees valley, is a broad fertile tract of agricultural land, for the most part below 300 feet in elevation, traversed by the middle and lower portions of most of the Yorkshire rivers, and stretching from the banks of the Tees to the borders of Nottinghamshire. Its light and sandy soils support ordinary lowland and woodland types of vegetation, the fauna partaking of the same character.

In former times much of the surface was frequented by the large wild animals, now classed amongst the extinct forms, and in particular the famous forest of Galtres, which stretched for many miles in extent from beneath the very walls of York, was inhabited by various beasts of chase, as wolves and red deer, and particu-larly famed as a harbour of wild boars. Parts of the district still remain to some extent in their pristine condition ; and such places as Pilmoor, and Strensall and Riccall Commons—the breeding places of the redshank, teal, snipe, blackheaded gull, &c., and some of them formerly of the ruff and other birds—with some boggy carrs and wet heaths of the North and East Ridings, and Askham Bog, still display their primitive characteristics. In the north-west the sylvan recesses of Hackfall still harbour the

badger, and at Hornby Castle near Catterick is to be found the
only decoy now existing in the county.   There is no lack of
woodland, especially towards the south, where at Edlington Wood
one of the last Yorkshire nests of the kite was taken, while that
of the hobby has been found at Rossington and in the woods at
Cawood, and in the latter, which were the largest in the county,
the raven and buzzard reared their young till within comparatively recent times.

In the extreme south the flat marsh-lands which lie between
the present and the old channels of the river Don, including the
carrs near Doncaster, and the famed levels of Hatfield Chase and
Thorne Waste, once ornithologically rich, even now present an
avifauna of considerable interest.   Formerly the three harriers,
the black-tailed godwit, and the ruff were among the species
breeding annually, and an island at the mouth of the Trent
afforded the last British nest and eggs of the avocet.   On Thorne
Waste was also the site of a small decoy fairly productive of
mallard, wigeon and teal, especially the latter.   This decoy, of
which no record is to be found, possessed three tubes, according
to Mr. H. W. T. Ellis, of Crowle, who has seen it in operation,
and states that it ceased to exist about forty years ago.   At the
present time Thorne Waste, which is about 6,000 acres in extent,
is the breeding haunt of the mallard, teal, redshank, black-
headed gull, and occasionally of the short eared owl and the
curlew.   On the intersecting drains the reed-warbler and species
of minor interest nest abundantly.

The Cleveland Hills, occupying the north-eastern portion
of the county, though inferior to the North-Western Fells in
extent and in elevation—reaching only to 1485 feet at Burton
Head—are no less picturesque and interesting.   Like them also
it is a region of high moorlands—frequented by red grouse and
twite, and in the spring and early summer by curlew and golden
plover, with, occasionally, a pair of stone-curlews, which here
find the northern limit of their breeding range in Britain—and
intersected by the ramified, well-wooded, and beautiful dales
drained by the Esk and by numerous branches of the Derwent.

The high lands of Cleveland present bold escarpments towards
the Tees valley and the central plain, and a lofty line of cliffs
towards the sea, reaching 680 feet in elevation at Boulby.   The
Howardian hills, below 520 feet in elevation, which separate the
vale of Pickering from the central plain, must be considered as
a southern spur or continuation of the Hambleton hills, as the
western escarpment of the Cleveland range is called.

The Cleveland fauna is highly interesting.   The badger is
more frequent here than elsewhere in the county, breeding in
several localities; there is reason to believe that the marten still
exists in small numbers in Eskdale, and the last Yorkshire wild
cat is known to have been killed on the Hambleton moors about
forty years ago.   The forest of Pickering once harboured the roe
deer, this district being the only part of Yorkshire in which this
animal is actually known to have existed in a wild state.   The raven
was formerly common in the district, and a pair are even now
observed in the vicinity of the coast.   This region was also the
resort of the hen harrier until 1850, to which date a few pairs
nested annually.   The short-eared owl has also on several occasions bred on the moors, and until quite recently a pair of pere-
grines nested annually.   On the moors the twite breeds sparingly,
and the curlew and golden plover not uncommonly.   The district
has on various occasions been visited by rare stragglers, such as
the pine grosbeak and Lapland bunting; while Tengmalm's owl
has occurred no less than three times.

Vale of Pickering.—South of the Cleveland hills is a small
tract of low-lying cultivated land, below a hundred feet in elevation, possessing a rich soil, and including a considerable extent of
carrs and low marsh-land.   This district, drained by the Derwent
and Rye, is shut in on all sides by high lands, and was, in all probability, formerly a lake, the outlet for its drainage even now being
at times inadequate, and in rainy seasons the lower portions are
liable to be flooded for miles in extent.   The chief zoological
interest of this small district is in the rich and varied fish-fauna
inhabiting the streams which converge at Malton—the barbel
being probably the only one deficient.   The Rye is famed for its
grayling and trout.

The Chalk Wolds.—A semi-circular range of rounded undulating chalk hills commences near the Humber at Ferriby, and sweeping first in a northerly and then in an easterly direction, terminates in a line of stupendous sea-cliffs at Flamborough Head.   Culminating at its north-west corner in Wilton Beacon, at an altitude of 805 feet, they present a bold front to the central plain on the west and to the vale of Pickering on the north, while by more gentle inclines their south-eastern or inner aspect merges into the low country of Holderness.

Originally a desolate, grassy, and stony sheepwalk — over which a horseman might ride for thirty miles at a stretch without meeting with a fence or other obstruction, and the resort of the great bustard and the stone-curlew—this district is now ranked amongst the best and most highly-farmed agricultural land of England.   The deeply excavated hollows in the Wolds are remarkable for the absence of streams, the only rivulets to which they give rise being the variable and intermittent ones called 'gypseys.' This deficiency of permanent streams decidedly affects the vertebrate fauna, probably accounting for the absence of such birds as the dipper, the sandpiper, and the grey wagtail, which occur and breed at corresponding altitudes amongst the hills of the north and west.   The characteristic fauna of the Wolds must now be regarded as a thing of the past.   The great bustard, which here found its northern limit in Britain, has long been driven out by cultivation, and the badger and the stone-curlew are on the verge of extinction, the chief bird now to be noted being the lapwing, which occurs in great abundance.

Holderness—a flat low-lying district of triangular outline interposed between the North Sea and the Humber, and separated from the rest of Yorkshire by the green Wold hills—is under an elevation of one hundred feet, with the exception of Dimlington Height, which is but one hundred and fifty-nine; and of all districts in the county is probably the one which has undergone the most decided physical transformation.   There can be little doubt that the aboriginal condition of the district, now rich and fertile corn-land, was that of a vast fen or swamp—the haunt of the bittern, the 'sholarde,' the crane, and the ruff, and possibly of the beaver. That it was originally fen is shown by the fact that in or before Haworth's time it was inhabited by characteristic marsh-loving insects, and even the swallow-tailed butterfly (*Papilio machaon*) is recorded as having formerly occurred. This is not improbable, for though the insect is now confined exclusively to the fens of Cambridgeshire and Norfolk, there is evidence to show that formerly it had a much more extensive range over England, even as far west as Shropshire, and southward to Dorsetshire and Hampshire. The sheets of water which formerly diversified the surface were made use of for the establishment of decoys for the capture of wild duck, and consequently we find that the greater number—four out of seven—of the decoys known to have existed in Yorkshire were here, at Home, Meaux, Watton, and Scorborough.

The impetus given to agriculture about the close of the last century, and the rapid development of high farming, proved fatal to much of the ornithological wealth of Holderness. The decoys were destroyed by the Holderness (1762) and the Beverley and Barmston (1800) drainage schemes; and many haunts were broken up by the general revival of agriculture.

Among the animals which once inhabited the district the herd of wild white cattle, which survived at Burton Constable till about the close of the last century, deserves mention.

Holderness, even now, is a rich ornithological district, the turtle dove and the quail being regular summer visitants, and the hawfinch breeds annually in some abundance. On Hornsea Mere—the largest natural sheet of water in Yorkshire—the reed warbler, the pochard, and the great crested grebe breed regularly; and it has produced some of the rarest Yorkshire visitants, such as the great white heron, the broad-billed sandpiper and others. The mere is inhabited by pike, which attain to enormous size, and are exceedingly destructive to the birds which frequent the water, especially the young ones, a circumstance probably explaining the absence of the little grebe.

The Yorkshire Coast-line—commencing at the mouth of the Tees, and extending 117 miles in length to Spurn Point—

is one of the most diversified possessed by any English county.

The estuary of the Tees—though by no means comparable in size or attractiveness to that of the Humber—is yet of considerable extent. It includes vast stretches of sands, which afforded the last breeding haunt of the seal in Yorkshire (one sandbank indeed bearing the name of 'Seal Sand'); also a series of low salt marshes bordered by sandhills, and intersected by pools and saltwater ditches—formerly the habitat of shore fishes, and an attractive resort for such migratory birds as the waders, ducks and geese. But, as so often has happened in the north of England, the development of trade has here sadly interfered with the natural productiveness of the district. The discovery of Cleveland ironstone—and consequent rapid rise of Middlesborough as a manufacturing and sea-port town—has involved a train of consequences which have done much to render the zoological riches of the Tees mouth almost a tale of the past. The navigation has been improved, foreshores embanked and reclaimed, docks and harbours built, breakwaters projected, and blast furnaces erected along the Coatham Marsh.

One of these furnaces, built within five hundred yards of the site of a decoy, caused—and no wonder—its discontinuance, about 1872. Formerly this decoy was fairly productive, and on one occasion yielded a haul estimated at five hundred. At any rate, so great was the number enclosed in the net, that it broke, and most of the ducks escaped, only ninety and nine being actually secured. Ducks are now but seldom seen on the Coatham Marsh, though the sheldrake nested on the sandhills as late as the year 1880, and may still continue to do so.

The Coatham Marshes and the adjoining Redcar coast possess an interest to the ichthyologist as the scene of the labours of Rudd and Ferguson, two of the most energetic observers that have worked at the fishes of the Yorkshire coast, and the results of whose researches are summarised in the lists appended to 'The Natural History of Redcar.' Many rare fishes have here been noted, including Ray's sea-bream (of which the first recorded or known specimen occurred here exactly 200 years ago, and was described by Ray and Willughby), the argentine or pearlside, the blackfish, and the Hebridal argentine.

The first ten or twelve miles of the Yorkshire coast, commencing from the mouth of the Tees, is low and fronted by a reach of firm sandy beach, but at Marske and Saltburn begins to rise. Here the Cleveland hills begin to present towards the sea a line of liassic and oolitic cliffs extending for forty-four miles, and terminating at the Castle Hill of Scarborough.   These Cleveland sea-cliffs—amongst the loftiest in England, and attaining their maximum height of 680 feet at Boulby—afford several breeding stations for the cormorant and the herring gull, whilst along their range the raven formerly bred in scattered pairs in suitable stations.   It is indeed probable that a single pair still lingers at a locality which it would hardly be politic further to indicate. The birds are there often seen, and as the species is not of a roving disposition, the probabilities are that they still nest.   The Scarborough Castle Hill—the outlying mass of rock which marks the southward termination of the Cleveland cliffs, was also in former times a breeding-station of this bird, and it is recorded to have nested there for the last time about 1850.

The coast—now the eastern termination of the Vale of Pickering—is comparatively low from Scarborough southward, and mostly composed of soft rocks which offer but slight resistance to the destructive action of the waves, save where the hard sandstone reef of Filey Brigg projects into the sea.   The shores are here composed of sandy beaches.   On the diluvial cliffs near Filey a few herring gulls breed annually.

Some distance S.S.E. of Filey the chalk deposits of England reach their northern termination in a lofty range of tide-washed mural precipices, the well-known cliffs of Speeton, Buckton, Bempton, and Flamborough, the most extensive and densely inhabited breeding resort of sea-fowl in England.   Here guillemots, puffins, razorbills, and kittiwakes breed in countless multitudes, the guillemots being by far the most numerous; and there are also a pair or two of herring gulls, and a few cormorants.   In a cave in Buckton cliff called 'The Cote' the rock dove breeds in great numbers, and its congener the stock dove is particularly numerous, breeding in the cliffs south of the North Landing at Flamborough.   The house martins have their nests under the ledges of the cliffs, and a few swifts in the crevices, whilst on the broken ground at the summit the rock pipit breeds somewhat commonly. Mingled with the sea-fowl breed innumerable starlings and jackdaws, and a pair or two of carrion crows nest annually, the sable hues of this bird, and of its congener the jackdaw, forming a striking contrast to the delicate plumage of the kittiwakes. The hooded crow has also occasionally remained to nest; and the highest portions of the cliffs are frequented by the peregrine, but, although the birds are always present in the season, it is uncertain whether they succeed in breeding annually, as one of them usually falls a victim to the gun.

The immense abundance of sea-fowl on these cliffs, and the ease with which they can be approached by means of boats, formerly led to their merciless slaughter for so-called sport and to supply the exigencies of fashion, and for years the locality was the scene of so much destruction that some of the species were at last utterly driven away, and others greatly diminished in number. This wanton cruelty was—as a matter of fact—the direct cause of the passing of the Sea Bird Preservation Act of 1869. The effects of that salutary measure have been most marked. The kittiwakes, which had become extremely scarce, are now quite numerous, and the cormorant, which had been entirely banished, has now again taken up its old breeding quarters, though as yet only to the extent of a pair or two. Among the species which formerly bred at Flamborough may be mentioned the shag and the raven, the former of which some few years prior to 1844 used to nest annually on the rocks, but now it breeds no nearer than the Farne Islands, and there only singly and irregularly.

The chalk cliffs attain their highest elevation of 436 feet at Buckton cliffs, declining thence eastward to 250 feet at the point of the headland, where the lighthouse is situated.

From its favourable geographical situation and bold physical aspect, the headland of Flamborough is famed as affording in the autumn a resting-place for many uncommon birds—such as the long-tailed duck, common, pomerine, and Richardson's skuas, the shearwaters, grebes, and petrels occurring annually.

But the famous headland and the great chalk cliffs with their ornithological wealth are not the only attractions which Flamborough possesses, for the base of the cliffs abounds with rock-pools, which, though as yet almost uninvestigated, will doubtless yield a varied and interesting fish-fauna.

The chalk terminates below Sewerby Hall, and is succeeded by the low diluvial cliffs and sandy beach of Bridlington Bay, stretching for forty-two miles in a bold concave sweep, which terminates in the marram-covered sandhills of Spurn. This line of coast, the eastern border of Holderness, composed of soft strata which are being steadily wasted away by the action of the sea, is comparatively uninteresting, and its vertebrate zoology offers but little that is worthy of special note until Spurn is reached. Spurn Point, the southern termination of the Yorkshire coast, is connected with the mainland of Holderness by a narrow neck of sandhills overgrown with marram-grass, a few yards in width, and preserved intact only by constant supervision, and at considerable expense. Were these intermitted the sea would speedily break through the isthmus and join the Humber, as it has done before now. Spurn is ornithologically rich. Birds migrating along the coast, or arriving from the east, find many temptations to linger. The miles of mudflats left bare on the Humber side of the isthmus by every receding tide offer great attractions and a never failing supply of food to various shore birds, and in the spring and autumn are frequented by great numbers of birds of this class. Many of these winter here—such as the bartailed godwit, grey plover, knot, turnstone, sanderling, and others. It is fortunate that Spurn is very strictly preserved, and equally so that this part of the coast is unsuitable for 'punting.' In winter thousands of duck and many brent geese are to be noted on the Humber; while woodcocks are sometimes observed in very great numbers on their arrival in the latter days of October.

The Geographical Position of Yorkshire, viewed from a faunistic standpoint, must be regarded as singularly favourable, as it presents a combination of advantages seldom equalled, both as regards the actual geographical range of the breeding species and the arrival of migrants and stragglers.

Situate about midway on the eastern seaboard of the British
Isles, and directly opposite the European continent, Yorkshire
is sufficiently far south to include species whose distribution is of
the southern type—such as the noctule, the nuthatch, and the
nightingale, which find in it the northern limit of their range,
while it is sufficiently far north to admit of the inclusion of such
species as the curlew, dunlin, &c., which here meet with their
southern breeding limits.

As regards the influx of migratory birds, a glance at the map
of Europe will at once show the advantageous position of the
county. Not only does its coast lie opposite that of the continent,
but Flamborough is on the same parallel of latitude as Heligoland,
the island which is so renowned for the myriads of migrants which
pass and repass it every spring and autumn. The observations
made there for many years by Mr. Gätke show that all the birds
passing over Heligoland do so in a direction due E. and W.
Such a line of flight, if sustained, would land the stream of immigrants upon the Yorkshire coast, and especially upon the prominent headland of Flamborough, which as a locality productive of
rare birds has few equals.

The configuration of the coast materially increases the advantage of the position, which is still more enhanced by the possession of two such points as Flamborough and Spurn. From the
Tees mouth the coast-line trends in a gracefully convex sweep in
a south-easterly direction to the headland of Flamborough—a
promontory which stands boldly out in the North Sea forty-three
miles in advance of the Tees mouth, and full fifty miles E. of
the mean longitude of the coast of Durham. South of Flamborough the coast-line recedes, and after the concave sweep of
Bridlington Bay, again advances, terminating in the long narrow
spit of Spurn, which—projecting sixty-two miles E. of the Tees
mouth—overlaps to a considerable extent the coast of Lincolnshire. Those birds—mostly waders and marine species—which
pursue a north and south course in their migrations, are in the
habit of following coast-lines, even though the latter keep well
out to sea. Such species making their way down the east coast
would probably pass the shores of Northumberland and Durham,

meeting with no obstruction till their progress is arrested by the promontory of Flamborough, where they are observed—and too often shot.   On leaving Flamborough they cross Bridlington Bay, and are either seen at Spurn, or, skirting Lincolnshire, pass on for the north coast of Norfolk—a well situated and rich ornithological county.

Flamborough and Spurn are by far the most favourable points for observing the arrivals of immigrants; and Spurn is considered far to surpass any portion of the Lincolnshire coast.   The tall cliffs of Cleveland probably offer attractions from their height and the secluded nature of the coast, but have never been systematically investigated by resident naturalists.   The winds which bring immigrant birds in the greatest numbers in the autumn are those not favourable to their passage.   When worn out by a long and adverse journey against contrary winds they drop on the first shore they reach, and the presence of woodcocks at Spurn and elsewhere on the coast depends on the prevalence of the strong N. or N.E. winds during their passage, which tire them out, and after which they are to be found on the point in great numbers.   On the contrary should the winds be light and favourable they simply pass on, dispersing themselves over the country in suitable situations, and very few would be observed on the coast.   On the Yorkshire coast the line of migration of all birds in the autumn is, as a rule, from E. to W., with sometimes a decided trend from points south of E., snow buntings and bramblings coming more from points north of E.   In the spring the warblers, swallows, &c., come from the S. and S.E., the line of migration of the cuckoo being from S.E. to N.W.

On theoretical grounds the geographical position of Norfolk, projecting as it does so prominently beyond the general coast-line of England, has usually been considered superior to that of any other county; but if the number of species be taken as a criterion, a comparison of the lists would show that practically Yorkshire is quite equal.   From the well-known fact that birds when migrating make for the most prominent and first-seen land, it is argued that the main stream would be attracted to the coast of Norfolk; but that coast being comparatively low, the probabilities are that the tall cliffs of Flamborough and Cleveland would compensate Yorkshire for any disadvantage caused by the less prominent outline of the county, and that the high lights of Whitby and Flamborough would be amongst the first seen by immigrants coming from the east.

The presence of rare American ducks and waders on the eastern shores of Britain has been accounted for by the supposition that they cross the Atlantic at high latitudes, and striking Norway follow the general trend of its coast-line. Reaching its southern termination they would endeavour to cross the sea in the same general direction—one which lands them in Norfolk or Yorkshire.

MAMMALIA.

An analysis of the list shows that of the seventy-two recognised British species forty-nine are recorded as occurring now or formerly in Yorkshire, including fourteen marine and thirty-five terrestrial forms.

Of the terrestrial species, three—the Wild Cat, the Roe Deer, and Wild White Cattle—are now extinct, though surviving in other parts of Britain. Four others—the Marten, Polecat, Badger, and Black Rat—which were formerly very abundant, have greatly decreased in number, and the first-named may be considered practically extinct. The Black Rat survives only in sea-port towns, where its numbers are to some extent kept up by importations. The Badger may—and probably will—long continue to linger in small numbers in the secluded dells among the oolitic rocks on the southern slopes of the Cleveland hills, where—though very local—it is not uncommon.

Three Bats—the Hairy-armed, Reddish-grey, and Whiskered—are reported from single localities only, and in the case of the two former upwards of forty years have elapsed since they were recorded.

The Fallow Deer is included in the list, but can hardly be considered as entitled to a place in the Yorkshire fauna, as it is known only in parks, and there is no evidence to show that it was ever truly wild. The Red Deer also is now only in parks, but its claim to be considered indigenous is valid, for there can be little doubt that those at Bolton and Wharncliffe are descended from the aboriginal wild stock once frequenting those districts.

The remainder of the terrestrial species do not call for much remark, save that the Noctule or Great Bat—and perhaps one or two other bats—find in Yorkshire the northern limit of their range in Britain. In stating the faunistic position of terrestrial mammalia, it has not been considered necessary to use the word 'resident,' inasmuch as all the species perforce come under that category.

Of the fourteen marine forms two are Seals, the remainder cetaceans. Both the Seals are now very rare casual visitants, although it is not a great number of years since one of them, the Common Seal, was quite an abundant resident at the Tees mouth. Of the twelve cetaceans the Porpoise only can be considered abundant, although the Grampus and possibly the Lesser Rorqual are of not uncommon occurrence ; the other species having occurred as stragglers only. No doubt other whales than those recorded visit the Yorkshire seas, but in the absence of evidence of their being examined by competent authorities the numerous 'finners,' 'grampuses,' and 'bottle-noses' reported in the newspapers from time to time must remain in obscurity.

The twenty-three British species which have not been found in Yorkshire include but one terrestrial form, the Varying or Mountain Hare, an animal which does not occur in Britain south of the Scottish highlands. The remainder are bats, seals, and cetaceans. Of the nine bats—a group which receives very scant attention and offers a wide field for research—there can be little doubt that other species remain to be discovered in Yorkshire. This is demonstrated by the fact that the present work is the means of adding an hitherto unrecorded form—the Whiskered Bat— to the Yorkshire fauna. Daubenton's Bat—which has indeed been reported, though not as yet fully proved to occur—is one which may be confidently expected as an addition ; and it is also quite within the bounds of possibility that one of the horse-shoe bats, which are considered by good authorities to perform an annual north and south migration, may yet turn up as a summer visitant. The four Seals are all species which from their rarity in England cannot be expected to occur; and most of the nine cetaceans can only be looked for as rare stragglers. But it is somewhat surprising that the Common Dolphin, which inhabits the south coast of England in more or less numbers, should not have been recorded as visiting the Yorkshire seas.

Comparisons of the mammalian fauna of Yorkshire with those of other districts for which lists have been published, show that the chief differences are in the bats and the marine forms. The terrestrial species, being nearly all of universal range in Britain, are present in all the lists. As the value of comparisons lies in their bearing upon questions of geographical distribution, it is interesting to find that the Dormouse is absent from the Norfolk fauna, and that, although present in Yorkshire and in Northumberland, its range does not extend into Scotland. That of four of the Yorkshire bats also—the Noctule, the Hairy-armed, Reddish-grey, and Whiskered Bats—falls short of Scotland, although the two latter are found in Northumberland and Durham. It is but of little use to pursue the comparison so far as regards the Seals and cetaceans, which usually rank as stragglers in all lists.

Regarding the extinct species, little need be said. The forms included in Mr. Harting's work on 'Extinct British Animals' are all that it is desirable to enumerate in such a work as this, being the only species which have survived in Britain to within the historic period, and whose former existence can be proved by other than palæontological evidence. Of these the Reindeer has never in historic times occurred south of Caithness: the evidence of the Beaver's existence in Yorkshire is entirely etymological, and that of the Brown Bear as entirely palæontological; while the former presence of the Wild Boar and the Wolf is attested by strong, reliable, and concurrent testimony.

The chief work which remains to be done for the Yorkshire terrestrial mammalia is to ascertain more completely the distribution—and especially the altitudinal range—of the smaller species, notably the Shrews, Mice, and Voles. The Lesser Shrew, Harvest Mouse, and Red Field-Vole, though all recorded, are scarcely known, and much overlooked.

BIRDS.

The avifauna of Yorkshire, compared with that of other counties, stands unrivalled, not only in its numerical extent, but also—a circumstance of much greater significance—in the inherent richness which is shown by the number of species breeding annually within its limits.

Excluding eleven species, which have been recorded on the strength of evidence more or less insufficient to establish their claims, the total number of birds on the Yorkshire list is 307. The Norfolk list given in the first volume of Stevenson's 'Birds of Norfolk' includes 291 species—to which must be added seven which have occurred in the county since that work was published, for the names of which we are indebted to Mr. Thos. Southwell, of Norwich, making a total of 298. The list given in Hancock's ' Birds of Northumberland and Durham,' published in 1874, comprises 268 species. But applying the same rules as are employed for the exclusion of doubtful species from the Yorkshire list, these totals are reduced to 290 for Norfolk, and 266 for Northumberland and Durham.

The species thus excluded from the Norfolk list are—the Pine-Grosbeak, Two-barred Crossbill, Mottled Owl, Red-breasted Goose, Harlequin Duck, King Eider, Hooded Merganser, all admitted on insufficient evidence, and the variety *Sabini* of the Double Snipe, to which Mr. Stevenson gives specific rank. Those omitted from the Northumberland and Durham catalogue are— the Purple Gallinule, probably an escape, and the Virginian Colin, an introduced bird. The species excluded from the Yorkshire list include one Casual and ten Accidental Visitants, whose names will be given when treating of those classes of birds.

The seven species which have been added to the Norfolk fauna during the past fifteen years are—White's Thrush, the Wall Creeper, the Lesser Grey Shrike, the Ortolan Bunting, the Golden Eagle, the Green-backed Gallinule, and the European Coal-titmouse.

A comparison of the three avifaunas—based upon a careful analysis in accordance with the faunistic definitions given on page xi—yields the following results:—

	North-umber-land and Durham	York-shire	Norfolk	
RESIDENTS - - - -	83	88	76	
SUMMER VISITANTS - -	30	32	31	
	113	120	107	= ANNUAL BREEDERS.
WINTER VISITANTS - -	43	47	50	
PERIODICAL VISITANTS-	12	17	26	
CASUAL VISITANTS - -	61	58	47	
ACCIDENTAL VISITANTS	37	65	60	
	266	307	290	= TOTAL AVIFAUNA.

This decided superiority of the avifauna of Yorkshire over those of the two maritime districts with which alone it is fair to institute comparisons, is to be accounted for by a combination of advantages. In Yorkshire the favourable geographical position of Norfolk is associated with its physical advantages and those of Northumberland and Durham, and when it is further considered that Yorkshire possesses in addition a much greater diversity of surface, soil, and climate than either, there remains little reason for surprise at the numerical excellence of its fauna. The superiority is not merely one of numerical extent. Casual and accidental visitants cannot be regarded as true members of any fauna, and the ornithological richness or poverty of a district can only be gauged by a comparison of the number of its residents and regular visitants, and more especially of that of the species which breed annually. In this respect too—as the table shows—the superiority of Yorkshire is well-marked, demonstrating still more forcibly the advantages possessed by the county which contains the greatest diversity of surface, a diversity ranging in this case from the low carr lands of the E.S.E. to the mountains of the W.N.W., with a coast-line affording both lofty and rugged cliffs and sandy flats, thus presenting every kind of habitat necessary for the presence of almost every type of bird which breeds in the British Isles.

The EIGHTY-EIGHT RESIDENT BIRDS include the following species which deserve special mention.

The Nuthatch, Hawfinch, Wood-Lark, Lesser Spotted Wood-pecker, Pochard, and Great Crested Grebe find in the county the northern limit of their distribution in Britain during the breeding season; though one or two of them have been known to nest occasionally or singly in districts still further north. The Curlew and Dunlin on the other hand find in it in like manner the limit of their southern range.

The Raven, Buzzard, and Peregrine Falcon, all formerly resident in some abundance, are now restricted to a few pairs of each species still breeding annually, the Buzzard, once so com-mon among the crags of the Yorkshire fells, being now the rarest of the three. The elegant little Goldfinch, too, is fast diminishing, and although widely distributed in the county is extremely local and nowhere numerous. The Sheldrake is one of the most local birds which nest in Yorkshire, only two breeding-haunts being known.

Yorkshire Heronries have greatly decreased during the present century. Formerly they existed in the following places, the date at which they ceased to do so where known being given in parenthesis: —Scorborough (1862 or 1864); Watton Abbey; Stork Hill, near Beverley; Hotham (1819); Swanland, near Hull; Sutton Wood, Sutton-on-Derwent, where in 1860 there were said to be a hundred nests; Hemsworth; Walton Park (1865); Scarthingwell; Bolton Woods; and Flasby, near Gargrave (1866). Those now in existence are enumerated at page 49.

The nesting of the Rock-Dove on inland cliffs, although given on good authority, is, it must be confessed, not perfectly satisfac-tory, as the bird so reported may possibly prove to be the Stock-Dove, a species which breeds not uncommonly in such situations.

Of THE THIRTY-TWO SUMMER VISITANTS the Nightingale, Reed-Warbler, Wryneck, Turtle-Dove, and Stone-Curlew reach in Yorkshire the northern limit of their annual distribution during the nesting-season. The Wryneck and Turtle-Dove have, however, been known in isolated instances to rear their young in localities further north.

That local and interesting bird the Pied Flycatcher is probably more abundant than in any other British county; its breeding-haunts being numerous and widely diffused.

The Lesser Tern has a single breeding-station, but it is a matter for regret that the species, fast decreasing in numbers, owing to persecution, bids fair to be classed among the extinct birds at no very distant period. It is somewhat singular that Yorkshire, having this species and the Ringed Plover in abundance, should not be able to include among its breeding birds the larger species of Terns and the Oyster-catcher; nor can it be ascertained that these birds ever did frequent its shores for such a purpose.

THE THIRTY-SEVEN WINTER VISITANTS do not include many species which merit special mention, but the following are uncommon birds of annual occurrence :—The Great Grey Shrike, Shore-Lark, and Rough-legged Buzzard. The last-named and some other birds of this class, as the Redwing, Hooded Crow, Crossbill, Short-eared Owl, and Tufted Duck, have in isolated instances been known to breed in the county.

None of THE SEVENTEEN PERIODICAL VISITANTS admit of much comment. The Pygmy Curlew, Common and Pomerine Skuas, are amongst the least numerous; while the Dotterel still visits the county annually as of old, but in gradually decreasing numbers. Its periodical visits are alluded to in the Northumberland Household Book, wherein (in 1512) it is set forth that 'at principal feestes . . . Dotterells to be bought for my Lord when thay ar in season and to be at jd. a pece.' Formerly the Osprey was of this class, and occurred regularly in the county every spring and autumn on its passage to and from its Scottish breeding-haunts, where it is now almost unknown.

Of THE FIFTY-EIGHT CASUAL VISITANTS the Dartford Warbler finds just within the extreme southern boundary of the county the most advanced of its northern outposts—at a locality in which on one occasion its nest and eggs were found.

The Bearded Reedling, Crested Titmouse, Ortolan Bunting, Chough, and Golden Eagle are noticeable as of exceptionally rare occurrence.

In addition to the fifty-eight, Richard's Pipit has been reported to occur, but upon evidence which cannot be admitted, from the lack of details necessary to substantiate its claim, although the species is one quite likely to occur.

In addition to the SIXTY-FIVE ACCIDENTAL VISITANTS included in the table at p. xxxiv, there are ten—the Purple Martin, Great Black Woodpecker, Hairy Woodpecker, Little Owl, Acadian Owl, Harlequin Duck, Passenger Pigeon (escape), Virginian Colin (introduction into Britain), Sooty Tern, and Laughing Gull—whose claims to a place in the Yorkshire fauna must be regarded as inadequate, though it is quite possible that further investigation may show some of them to have been genuine occurrences.

Four species which have occurred in Yorkshire—the Mottled Owl, Lesser Kestrel, Cuneate-tailed Gull, and Bulwer's Petrel—have not been known to visit any other British locality, and the one last named has not even occurred elsewhere in Europe.

As to the occurrence of the Mottled Owl, there is no reason to doubt its validity, for Dr. Hobson thoroughly sifted the evidence at the time.

Regarding the Lesser Kestrel—another species whose occurrence has been challenged—the writers are perfectly convinced, from their personal acquaintance with the gentleman who obtained it, that it was a genuine one. Mr. Harrison shot the bird solely on account of its diminutive size, and after he had observed it about his residence for some days. It has been suggested that Graham, of York, to whom the specimen was taken for preservation, substituted for it a foreign skin; but Mr. Harrison, whose attention was particularly impressed by the bird, and who is, moreover, a good ornithologist, could hardly have been imposed upon in such a manner. The geographical range of the species would not preclude its visiting Britain, and it has been known to occur at Heligoland, immediately opposite the Yorkshire coast. The time of year has also been urged as an argument against the validity of the occurrence, but it may be pointed out that the Hobby—usually regarded as a summer visitant—has on several occasions been taken in Yorkshire in mid-winter.

The Cuneate-tailed Gull—of which not half-a-dozen specimens are known to exist—is specially interesting from its extreme rarity; and although there is a discrepancy of dates in the two versions published at the time of the occurrence, there is no ground for doubting its genuineness.

Amongst the species which have occurred in Yorkshire of which very few British examples are known, may be cited the Rock-Thrush, Orphean Warbler, White-spotted Bluethroat, Tawny Pipit, Pine-Grosbeak, White-winged Crossbill, Red-winged Starling, Eagle Owl, Greenland and Iceland Falcons, Swallow-tailed Kite, American Bittern, Red-breasted Goose, Polish Swan, Ruddy Sheldrake, Buffel-headed Duck, King Eider, Steller's Duck, Barbary Partridge, Andalusian Hemipode, Yellowshank, Broad-billed Sandpiper, Gull-billed Tern, and White-winged Black Tern.

Treating of Yorkshire birds generally, it may be remarked that many resident birds are to a greater or less extent migratory, shifting their quarters from one locality to another according to the season, as for example the Curlew, which breeds on the high moors in the summer and retires to the shores during the winter, while the Thrush and Pied Wagtail remain through that season in much reduced numbers. With some species, as the Long-eared Owl, Kestrel, and Reed-Bunting, it is very possible that the individuals found in summer migrate southwards for the winter, and are replaced by others from localities still further north. Such a circumstance, however, would not in the least militate against the claim of the species to be considered as resident.

On the other hand, there are species—true winter visitants, though ranked in some county lists as residents—of which a few are found in the district throughout the year, but they cannot be regarded as 'residents' in the true sense of the term, for the individuals remaining through the summer are immature and non-breeding birds. These remarks are applicable to (amongst other species) the Turnstone (of which about a score remain at Spurn throughout the summer), the Dunlin, Common Scoter, Common and Lesser Black-backed Gulls, and Red-throated Diver, all of which are to be found in more or less numbers on or off the coast at all seasons. The fact of individuals remaining in this way is but an exceptional one, not affecting the faunistic location of the species.

In addition to the species at the present time regularly breeding in the county, others must be mentioned as having formerly nested annually, but which are now entirely banished in consequence of persecution, or of the great changes wrought in their former haunts; and instead of being claimable as members of the two classes which furnish the breeding species, they can now only be ranked as Casual or as Accidental Visitants, of more or less rare occurrence.

Such species include the Kite, which there can be no doubt was once very abundant, but of whose breeding the information is so meagre that only two actual instances can be cited. The three Harriers, though local, were once fairly abundant, the Hen-Harrier, perhaps, being the least so, though it is now the most frequent as a casual visitant. The Marsh-Harrier, on the contrary, is now one of the rarest, whilst Montagu's Harrier was the most widely distributed and the last to linger on the Yorkshire heaths. The Hobby, earlier in the present century, was regarded as far from uncommon in South Yorkshire, but it is now seldom seen, and only three instances of its breeding in the county can be cited. Although the Bittern was formerly abundant, and doubtless bred in the county, there is no positive record in existence of a nest or eggs having been found. Regarding the Bustard, which formerly had its most northern residence in Britain on the wolds of Eastern Yorkshire, all the information obtainable has been amassed. In this case, a justifiable departure from the general plan of the work has been made, in order to place on record—ere it is lost for ever—all the information which it is possible to obtain of its former existence. The Shag, though now quite unknown even as a casual breeder, once nested in some abundance on the cliffs at Flamborough. It is satisfactory to have information so interesting on the high authority of Mr. Arthur Strickland. The former breeding of such birds as the Grey-lag Goose, Avocet, Ruff, Black-tailed Godwit, and Black Tern, is mentioned under the head of the respective species in the catalogue.

Such are the principal losses which Yorkshire has sustained in breeding-birds, the result chiefly of the changes which have taken place in the physical aspect of the county.

To all rules there are, of course, exceptions; and it is therefore not surprising to find that winter visitants, like the Short-eared Owl, Hooded Crow, Redwing, Tufted Duck, and Rough-legged Buzzard; casual visitants, as the Dartford Warbler, Bearded Reedling, Siskin, Crossbill, Cirl Bunting, and Redbacked Shrike; and even accidental visitants, as the Orphean Warbler, have occasionally and in isolated instances remained to breed.

With respect to migratory species, the dates of arrival and departure quoted in the catalogue are those observed on the coast, as more likely to be reliable than observations made in inland localities.

It is of interest to note that in addition to the Mottled Owl, Lesser Kestrel, Cuneate-tailed Gull, and Bulwer's Petrel—which are unique as British specimens—there are several other species whose first occurrence in Britain was in this county. Amongst these are the Waxwing (1681), Red-breasted Goose (1766, one also occurring near London about the same time), Scops Owl (1805), Red-legged Falcon (April, 1830), and Orphean Warbler (1849); and probably also the Eagle Owl, mentioned by Pennant in 1768 as having once been shot in Yorkshire.

The species recorded in the present work for the first time as Yorkshire birds are the Lapland Bunting, Dartford Warbler, Ruddy Sheldrake, Broad-billed Sandpiper, Black-winged Stilt, and Wilson's Petrel. It may also be remarked of Tengmalm's Owl, that out of the seventeen known British specimens no less than five have occurred in this county, and three of these Mr. Clarke has had the personal pleasure of adding on the most unquestionable authority. In addition to this, numerous occurrences of rare species—which have hitherto remained unpublished—now appear for the first time in print.

REPTILES AND AMPHIBIANS.

The British lists given include nine reptiles and seven amphibians, six of each class being recorded as occurring in Yorkshire. The three British reptiles not hitherto found in the county are the Smooth Snake, a species found only in the New Forest, the Green Lizard, a questionable native of Britain, and the Sand-Lizard, a species not unlikely to inhabit Yorkshire, for which indeed it has been reported, but not sufficiently authoritatively to warrant its being accepted. It is notable that both the Turtles—which are however but accidental stragglers from tropical seas—have occurred off the Yorkshire coast.

Of the amphibians all the species occur in Yorkshire except the Edible Frog, only admissible as British on the strength of examples introduced within living memory, and successfully naturalised in the Cambridgeshire and Norfolk fens. The most interesting of the Yorkshire amphibians is the Natterjack Toad, which the present work is the means of adding to the fauna. The distribution of the Newts being imperfectly worked out requires further investigation, when no doubt the range of the Palmated Newt will be found more general than at present appears to be the case.

Compared with other county-lists, very little difference is observable. Norfolk possesses the Edible Frog, and Northumberland and Durham the Sand-Lizard, while both are deficient in the two Turtles.

FISHES.

This group has been but imperfectly investigated in Yorkshire, very few naturalists having devoted attention to the study of the marine species; consequently there are but scanty materials for ascertaining the faunistic position, or even the relative abundance or scarcity of most of the species. Nevertheless the knowledge already acquired justifies the opinion that Yorkshire possesses a rich and varied pisci-fauna, accounted for by the diversity which exists in the sea-bed, and partly also by the proximity of the 'Dogger Bank,' that famous nursery of the North Sea.

Of the 249 British species of fishes, 155 are recorded as having occurred in Yorkshire. The claims of seven—the Basking Shark, Large-spotted Dogfish, Starry Ray, Northern Chimæra, Sparus or Dentex, Four-horned Bullhead, and Atherine—are however insufficient, until confirmed by further research, to entitle them to admission into the county fauna.

Of the 148 species thus recognised 32 are inhabitants of fresh and 4 of estuarine or brackish waters, the remainder being marine.

Of the fresh-water fishes none call for remark save the Grayling, a form interesting on account of the irregularity of its distribution in Britain, which is probably more widely diffused and abundant than in other counties.

Of the marine forms several species are worthy of note. The Argentine, Ray's Sea-Bream, and Banks' Oarfish appear to have occurred more frequently than elsewhere in Britain. The Eagle Ray and the Hebridal Argentine, of which very few British examples are known, have occurred singly in Yorkshire. Other rarities noted are the Fox Shark, Greenland Shark, Spinous Shark, Black Sea-Bream, Bergylt, Maigre, Blackfish, Spanish Mackerel, Tunny, Red Bandfish, Jago's Goldsinny, Tadpole Hake, Sea-horse or Hippocampus, Sunfish and Lancelet.

Several species have been first described or recorded as British from specimens taken in the county, such as Banks' Oarfish, Ray's Sea-Bream, and the Short-headed Salmon, while Yorkshire specimens of other species have been figured by Yarrell, or represent them in the British Museum collections.

The present work is the means of adding the following species to the county fauna—the Sharp-nosed Ray, Long-nosed Skate, Striped Surmullet, Black Sea-Bream, Red Bandfish, Striped Wrasse, Goldsinny, Poor or Power-Cod, Rough Dab or Sandsucker, Sailfluke, Smear Dab, and Dwarf Sole—twelve in number.

Dr. Lowe's useful list of the Fishes of Norfolk, published in 1873, includes 124 of the species here admitted as British, besides other forms not allowed to be such. Eight of the Norfolk fishes have not occurred in this county, and 33 Yorkshire fishes not in Norfolk.

The Fisheries of the Yorkshire coast are important, and produce most of the British food-fishes, the Gadidæ being probably the most valuable. The coast itself, with its rough rocky ground, is unfavourable for trawling inshore, that method of fishing being mostly carried on in the North Sea and on the Dogger Bank, whilst what is termed the long-line fishing is conducted about ten or twenty miles from the coast. Formerly stake-nets were worked in Bridlington Bay, which produced Salmon, Salmon-trout and Sturgeon, and sometimes rarities like the Swordfish were taken; but these have now been discontinued.

The only previous enumeration of Yorkshire fishes as a whole was that which Mr. Meynell read in 1844 to the British Association, a brief abstract only of which was printed in the report. This list—which was stated to contain 140 species—does not appear to have ever been published, and our endeavours to trace the existence of the manuscript have been unsuccessful. This is to be regretted, as there can be little doubt that much information has thus been lost.

There yet remains much work to be done in the study of the marine fishes of the Yorkshire coast, chiefly the ascertainment of their faunistic position, migratory movements, and relative abundance. Particular attention should be directed to the collection of littoral and rock-pool species, such as the blennies, gobies, pipe-fish, and other small forms. The crab and lobster 'pots' are attractive to many such, which may be thus easily obtained. The varieties of the Common Stickleback also require investigation, and also the various species of Grey Mullets which have been differentiated within recent years.

To increase the usefulness of the fish-list, a departure has been made from the general plan of the work, by the inclusion of the various local names used, so far as we are able to give them.

A General Summary of the Vertebrate Fauna of Yorkshire and of the British Isles, may thus be expressed :—

	Yorkshire.	British Isles.
MAMMALIA :		
TERRESTRIAL - -	32	45
MARINE - - -	14	27
BIRDS - - - - - -	307	380
REPTILES :		
TERRESTRIAL - -	4	7
MARINE - - -	2	2
AMPHIBIA - - - -	6	7
FISHES :		
FRESHWATER - -	32	53
MARINE - - -	116	196
	513	717

Few words are needed in concluding the introductory observations. The plan of the work, as defined in the opening
paragraph of the introduction, has been rigidly adhered to, the
necessary information as to the faunistic position, migrations, and
geographical range of the various species being expressed with
studied brevity, irrelevant matter carefully excluded, and literary
style invariably subordinated to scientific accuracy. The only
deviation from the rule of brevity is in the case of species like
the Great Bustard and Common Seal, which have become extinct
so recently as to impose upon naturalists the moral obligation to
search out all the traditions while they are fresh in the memories
of men, and to rescue from oblivion every possible scrap of
evidence.

In the case of rare species in the public museums of the
county, and also in a few of the instances of their being in
private collections, it is deemed useful to indicate where they are
now preserved.

The authors offer no apology for undertaking the task, for none is necessary. A Yorkshire list had never been published, and it was within their power to supply the desideratum, partly because for some years they had systematically investigated the literature of the subject, and partly on account of the advantage they possessed in being in communication with the leading naturalists of the county. On another page they have endeavoured to express—even though inadequately—their sense of obligation to those who have contributed towards the materials upon which this work is founded. That such a work as the present was required is amply proved by the large amount of interesting matter now published for the first time. The voluminous nature of the evidence upon which the work is founded is shown by the list of contributors given at the end ; and the separate authorship of each section of the list is sufficiently indicated by the flysheets inserted.

It cannot of course be expected that error has been entirely avoided, especially in the case of a work which in the essential part of its design has no predecessor but Mr. J. E. Harting's excellent and indispensable 'Handbook of British Birds'; but —so far as an exhaustive examination of the literature of the subject, together with the unanimous and generous assistance received from the naturalists of the county, and the exercise of scrupulous care in judging evidence, are concerned—it is believed that the chances of error have been reduced to a minimum.

EXTRACTS FROM ANCIENT RECORDS.

A LTHOUGH Yorkshire does not possess evidence so definite or so
complete as that which exists for some other counties, there are records
extant which in some degree serve to throw light on the fauna which once
inhabited the district. A brief summary of the animals mentioned in three
of the principal documents will not be devoid of interest.

In 1466, as Leland writes in his 'Collectanea,' a great feast was given in
the archiepiscopal palace at Cawood, on the occasion of the 'intronization'
of 'George Nevell, Archbishop of York, and Chauncelour of Englande, in
the vj. yere of the raigne of Kyng Edwarde the fourth'; the goodly provision
made for which included

'. . Wylde Bulles, vj.; . . . Swannes, CCCC.; Geese, MM.;
. . . Plovers, iiii.C.; Quayles, C. dosen.; Of the fowles called Rees, CC.
dosen.; In Peacockes, Ciiii.; Mallardes and Teales, iiii.M.; In Cranes,
CC.iiii.; . . . . Pigeons, iiii.M.; Conyes, iiii.M.; In Bittors, CC.iiii.;
Heronshawes, iiii.C.; Fessauntes, CC.; Partriges, v.C.; Woodcockes, iiii.C.;
Curlewes, C.; Egrittes, M.; Stagges, Buckes, and Roes, v.C. and mo.; Pykes
and Breames, vi.C. and viii.; Porposes and Seales, xii.'

The fish dinners included 'Red Herrynges ; Salt fysch ; Luce salt ; Salt ele ;
Kelyng, Codlyng and Hadocke . . .; Thirlepoole . . .; Pyke in
Harblet ; Eles baked ; Samon chynes . .; Freshe Salmon jowles ;
Salt Sturgion ; Whytynges ; Pylchers ; . . Makerels ; Places . .;
Barbelles ; Conger . .; Troute ; Lamprey . .; Bret ; Turbut ;
Roches ; Lynge . .; Tench . .; freshe Sturgion ; Great Eeles ;
. . . Cheuens ; Breames ; Rudes ; Lamprones ; Small Perches . .;
Smeltes . .; Small Menewes ;' besides 'crabbes' and 'lopster.'

The document next in point of age dates 1512 and is entitled 'The
Regulations and Establishment of the Household of Henry Algernon Percy,
the Fifth Earl of Northumberland, at his castles of Wresill and Lekinfield in Yorkshire. Begun Anno domini M.D.XII.' This valuable book,
usually called the 'Northumberland Household Book,' well shows the almost
regal state maintained by the Percys, and no doubt other great nobles, in the
time of Henry the Eighth.

Therein we find that while 'chekyns' cost a halfpenny each and 'hennys'
2d. each, it was 'thought good that no pluvers be bought at noo Season bot

oonely in Chrystynmas and princypall Feestes and my Lorde to be servyde therewith and his Boordend and non other and to be boght for j*d*. a pece or j*d*. ob. at moste.' Then it was thought good that ' my Lordes Swannys ' be taken and none bought ' seynge that my Lorde hathe Swannys inew of hys owne.' Other birds were to be bought for ' my Lordes owne Mees ' and the prices are duly set forth. ' Cranys ' were to be at ' xvj*d*. a pece,' ' Hearon-sewys ' at ' xij*d*.,' ' Mallardes ' at 'ij*d*.' ' Teylles ' at ' j*d*.' ' bot if so be that other Wyldefowll cannot be gottyn,' ' Woodcokes ' at ' j*d*. a pece or j*d*. ob. at moste,' ' Wypes ' at ' j*d*.,' 'Seegulles ' ' so they be good and in season and at j*d*. a pece or j*d*. ob. at the moste,' ' Styntes ' ' so they be after vj a j*d*.,' ' Quaylles ' at ' ij*d*.,' ' Snypes ' ' after iii a j*d*.,' ' Pertryges ' at ' ij*d*. a pece yff they be goode,' ' Rede-shankes ' after ' j*d*. ob. the pece,' ' Bytters ' at ' xij*d*. a pece so they be good,' ' Fesauntes ' at ' xij*d*.,' ' Reys ' at ' ij*d*.,' ' Sholardes ' at ' vj*d*.,' ' Kyrlewes ' at ' xij*d*.' ' Pacokes ' at ' xij*d*. a pece and noo Payhennys to be bought,' ' See-pyes ' have no price allotted, ' Wegions were to be at ' j*d*. ob.,' ' Knottes ' at ' j*d*.,' ' Dottrells ' ' when thay ar in Season and to be at j*d*. a pece,' ' Bustardes ' with no price affixed, ' Ternes ' ' after iiij a j*d*.' ' Great Byrdes after iiij a j*d*.' ' Smale Byrdes ' ' after xij a j*d*.,' and ' Larkys ' ' after xij for ij*d*.' There is an interesting memorandum that ' it is thought good that all manner of Wyld-fewyll be bought at the fyrst hand where they be gotten and a Cator to be apoynted for the same For it is thought that the pulters of Hemmyngburghe and Clyf hathe great advantage of my Lorde Yerely of Sellynge of Cunys and Wyldefewyll.' As to fish it is provided ' that a Direccion be taken at Lekyngfeld with the Cator of the See what he shall have for every seam of Fysch thorowt the Yere to serve my Lordes hous :' also ' that a Direccion be taken with my Lordes Tenauntes of Hergham and to be at a serteyn with theme that they shall serve my Lordes hous thrugheowt the Yere of all manar of Fresh Water Fysche.' Copies of warrants too long to be quoted are given, one for ' Twentie Signettes To be takenne of the breide of my Swannys within my Carre of Arromme within my Lordeschip of Lekinfeld ' for the Christmas feast of 1514.

It is of interest to find ' An Account of all the Deer in the Parks and Forests in the North belonging to the Earl of Northumberland, taken in the 4th Year of Henry VIII. Anno 1512 ' by which it seems that there were

' In Yorkshire.

Topcliff Great Park	...	Fallow-Deer	...	558
Topcliff Little Park	...	Ditto	...	291
Spofforth Park ...	...	Ditto	...	180
Spofforth Wood	...	Ditto	...	43
Wresill Park	...	{ Red Deer, 42 } { Fallow, 92 }	...	135
Wressel Litle Park	...	Fallow	...	37
Newsham Park ...	...	Ditto	...	324
Leckinfield Park	...	Ditto	...	249
Catton Park	...	Ditto	...	79.'

In the year 1526 another member of the great family of the Neviles, Sir John Nevile of Chevet near Wakefield, High Sheriff of Yorkshire, gave a banquet to celebrate his daughter's marriage.   Some of the charges are thus given:—'two roes, 10s., and for servants going, 15s.; swans, 15s.; nine cranes, 1l. 10s.; twelve peacocks, 16s.; six great pike, for flesh dinner, 10s.; 21 dozen conies, 5l. 5s.; three venison, red deer hinds, and fetching them, 10s.; twelve fallow deer, does, . . .; thirty dozen mallards and teal, 3l. 11s. 8d.; two dozen heron-sewes, 1l. 4s.; twelve bitterns, 16s.; eighteen pheasants, 1l. 4s.; forty partridges, 6s. 8d.; eighteen curlews, 1l. 4s.; three dozen plovers, 5s.; five dozen stints, 9s.; sturgeon on goile, 5s.; one seal, 13s. 4d.; one porpoise, 13s. 4d.'

In 1530, another daughter was married, and the expenses are returned pretty much as before.   The prices were, for swans, 6s. each; cranes, 3s. 4d. each; heronsews. 12d.; bytters, 14d.; and conies, every couple 5d.

In 1528, Sir John acted as Sheriff, and returned his charges as follows :—
'Item, two barrells of herrings, 1l. 5s. 6d.; Item, two barrells of Salmon, 3l. 1s.; . . . Item, in great pike, and pickering. received of Rither, 6 score, 8l.; Item, 12 great pike from Ramsay, 2l.; Item. in pickerings from Holderness, iiiicc., 3l.; Item, received of the said Rither 20 great breames, 1l.; Item, received of the said Rither 12 great tenches, 16s.; Item, received of the said Rither, 12 great eells and one hundred and six jowling eels, and 200 brewitt eels, and twenty great rudds, 2l.; Item, in great fresh salmon, 28, 3l. 16s. 8d.; Item, a barrell of sturgeon, 2l. 6s. 8d.; Item, a firkin of seal, 16s. 8d.; . . . Item, three bretts, 12s.; . . . Item, a draught of fish, 2 great pikes and 200 breams, 7l. 6d. 8d.'   Such was the fare provided at the Lent Assizes.   At the Lammas Assizes in the same year flesh was provided at about the same prices as above cited for the wedding feasts, with the addition of ' 12 shovelards, 12s.; item, 10 bytters, 13s. 4d.; item, 80 part-ridges, 1l. 6s. 8d.; item, 12 ffesants, 1l.; item, 20 curlews, 1l. 6s. 8d.; item, curlew knaves, 32, 1l. 12s.; 6 dozen plovers, 12s.; item, 30 dozen pigeons, 7s. 6d.,' &c., showing a slight advance on the prices of 1526.   At the same Lammas assizes the fish was charged as follows :—' First three couple of great ling, 12s.; Item, 70 couples of heberdines, 2l.; Item, salt salmon, 1l.; Item, fresh salmon and great, 3l. 6s. 8d.; Item, 6 great pikes, 12s.; Item, 80 pickerings, 4l.; Item. 300 great breams, 15l.; Item, 40 tenches, 1l. 6s. 8d.; Item, 80 jowling eels, and brevet eels, and 15 rudds, 1l. 12s.; Item, a firkin of sturgeon, 16s.; Item, in fresh seals, 13s. 4d.; Item, Eight Seam of fresh fish, 4l.; Item, 2 bretts, 8s.

The accounts of the Cliffords, of Skipton Castle, as given in Whitaker's Craven, afford evidence of the existence nearly three centuries ago of the same fish in Malham Tarn as now inhabit it.   In 1606 there is an entry of 2s. 6d. 'P'd to H. H. being at Mawater, watching the well-head for stealing the trouts coming unto this Kitt Time,' and in 1609 was paid ' For getting 33 pearch and troot from Mawater for my lo. and judge, iis. vid.'

MAMMALIA.

WM. DENISON ROEBUCK.

B

Class 1. MAMMALIA.

Sub-class *MONODELPHIA*.

Order **CHIROPTERA**.

Sub-order *MICROCHIROPTERA*.

Family **RHINOLOPHIDÆ**.

1. **Rhinolophus hipposideros** (*Bechst.*). **Lesser Horseshoe Bat.**

2. **Rhinolophus ferrum-equinum** (*Schreb.*). **Greater Horseshoe Bat.**

   The occurrence of either of the Horse-shoe Bats in Yorkshire has never yet been authenticated, although both have been occasionally reported : such specimens as I have been able to examine have proved to be of some other and commoner species.

Fam. **VESPERTILIONIDÆ**.

3. **Synotus barbastellus** (*Schreb.*). **Barbastelle.**

4. **Plecotus auritus** (*L.*). **Long-eared Bat.**

   Generally distributed, common. In some localities is more numerous than the pipistrelle.

5. **Vesperugo serotinus** (*Schreb.*). **Serotine.**

6. **Vesperugo discolor** (*Natterer*). **Particoloured Bat.**

7. **Vesperugo noctula** (*Schreb.*). **Noctule.**

   Widely distributed and not uncommon, becoming less numerous in the north and west. Found at 700 feet elevation at Carperby in Wensleydale. Its range in Britain does not appear to extend further north than Yorkshire.

8. **Vesperugo leisleri** (*Kuhl*).   **Hairy-armed Bat.**
Only one occurrence. Three specimens were obtained by
Mr. F. Bond about 40 years ago, which had been taken
from an old factory chimney-shaft at Hunslet, near Leeds;
one of them, a male, is still in his collection (Bond, MS.).

9. **Vesperugo pipistrellus** (*Schreb.*).   **Pipistrelle.**
Generally distributed, abundant.

10. **Vespertilio dasycneme** *Boie.*

11. **Vespertilio daubentonii** *Leisler.*   **Daubenton's Bat.**

12. **Vespertilio nattereri** *Kuhl.*   **Reddish-grey Bat.**
Only once recorded. A pair were taken alive in June, 1840,
out of an old tree in Oakwell Wood, Birstal (Denny, Ann.
and Mag. Nat. Hist., Aug. 1840, p. 385).

13. **Vespertilio bechsteinii** *Leisl.*   **Bechstein's Bat.**

14. **Vespertilio murinus** *Schreb.*   **Mouse-coloured Bat.**

15. **Vespertilio mystacinus** *Leisl.*   **Whiskered Bat.**
Reported from Great Mytton only. I have the satisfaction
of adding this bat to the Yorkshire fauna, Mr. F. S.
Mitchell, of Clitheroe, having sent me for determination a
specimen taken in the church at Great Mytton, a village
close to the confluence of the Ribble and Hodder.

Order INSECTIVORA.

Fam. ERINACEIDÆ.

16. **Erinaceus Europæus** *L.*   **Hedgehog.**
Universally distributed and abundant, ascending to 1300
feet or more.

Fam. TALPIDÆ.

17. **Talpa europæa** *L.*   **Mole.**
Universally distributed and very abundant; ascends to the
summits of the highest mountains, such as Whernside and
Ingleborough.
White and cream-coloured varieties, though of rare occur-
rence, have been reported from various localities; and at
Stillingfleet, near York, white moles appear to be of some-
what persistent occurrence.

Fam. SORICIDÆ.

18. **Sorex tetragonurus** *Herman.* **Common Shrew.**
Generally distributed, abundant. Ascends to 1300 feet.

19. **Sorex minutus** *L.* **Lesser Shrew.**
Reported from widely separated localities, and probably not uncommon, but is seldom distinguished from the Common Shrew.

20. **Crossopus fodiens** (*Pallas*). **Water Shrew.**
Generally distributed but not very numerous; commoner in some districts than others.

Order **CARNIVORA.**

Sub-order *FISSIPEDIA.*

Section ÆLUROIDEA.

Fam. **FELIDÆ.**

21. **Felis catus** *L.* **Wild Cat.**
Extinct, the Hambleton Hills having been its final refuge in Yorkshire. The last specimen there was trapped by my friend Mr. John Harrison, on his farm at Murton, near Hawnby, in the winter, about 1840. Other testimony confirms the opinion that the Hambleton Hills were the wild cat's latest haunt. There is no proof that it ever inhabited the Fells of the north-west, though in all probability it once existed there. The evidence of its former existence in South Yorkshire is confined to entries in the churchwardens' accounts at Ecclesfield, of sums paid in 1589 and 1626 for the destruction of 'wylde catts'; and to a legend of doubtful origin, of an encounter—fatal to both—between a wild cat and a man of the family of Cresacre, at Barnborough.

Section CYNOIDEA.

Fam. **CANIDÆ.**

Canis lupus *L.* **Wolf.**
Extinct, formerly abundant. There is conclusive evidence of various kinds to show, not only that this animal occurred, but that in former times it was abundant in the

county; and there is good reason to believe that it lingered longer in Yorkshire than elsewhere in England, and as late as the reign of Henry VII. There is documentary evidence of the existence of wolves at Flixton-on-the-Wolds in the time of Athelstan, at Roche Abbey in 1186, at Bolton in Wharfedale in 1306, and at Whitby in 1369. Evidence less precise but equally credible proves that they inhabited Knaresborough and Galtres Forests; Langwith, near York; Marske, in Swaledale; and various places in South Yorkshire, as Woolley, Dodworth, Silkstone, Aughton, Ulley, and Slade Hooton. Local legends in which the wolf plays a prominent part are related for Sittenham, in the Forest of Galtres; for John o' Gaunt's Inn, near Rothwell; and for Howley Hall, near Batley. Bones have been found in caves at Kirkdale and Dowka-bottom, as well as in river deposits.

22. **Canis vulpes** *L.* **Fox.**

Generally distributed and abundant, though less numerous on the western or hilly side of the county. Religiously preserved for purposes of sport, numerous packs of hounds, some of them dating back for centuries, being maintained.

Section ARCTOIDEA.

Fam. MUSTELIDÆ.

23. **Martes sylvestris** *Nilss.* **Marten.**

Extremely scarce, and restricted to one or two localities; formerly abundant, and generally distributed. The decrease in its numbers appears to have been comparatively rapid; the evidence in my possession concurs in showing that about the commencement of the present century the marten was common in many districts; while during the past thirty years its occurrence has been quite exceptional and unlooked for. The only instances in which it has occurred of late are:—Lees Head, near Whitby, one, fifteen or twenty years ago (Stephenson, MS.), and another in 1877 (Land and Water, p. 224); Cannon Hall Park, Barnsley, about 1878 (T. Lister, MS.); and Buckden, Wharfedale, winter of 1880 (Bishop, MS.).

[**Martes foina** (*Erxl.*). **The Pine Marten** has been removed from the British list by the late Mr. E. R. Alston, who considered that all British-killed martens are referable to *M. sylvestris.*]

24. **Mustela vulgaris** *Erxl.* **Common Weasel.**

Universally distributed, abundant everywhere.

25. **Mustela erminea** *L.* **Stoat. Ermine.**

Universally distributed, but not so numerous a species as the weasel. In the north-western fells this animal is known as the 'Polecat.'

26. **Mustela putoria** *L.* **Polecat. Foumart.**

Irregularly distributed, extremely rare, and fast becoming extinct. Half a century ago this species was generally abundant. Escaped ferrets are not unfrequently mistaken for this animal.

27. **Lutra vulgaris** *Erxl.* **Otter.**

Occurs in limited numbers in all the rivers, with the exception of the polluted streams of the manufacturing districts. Apparently absent also from Holderness.

28. **Meles taxus** (*Schreb.*). **Badger.**

Very local, and extremely limited in numbers. Its present haunts seem to be restricted to calcareous formations, which afford it suitable habitats. In Cleveland it still breeds near Pickering, near Kirby Moorside, at Hovingham, and at Hackness. On the Wolds it appears only to occur at Sledmere and Hunmanby, where, however, it is now all but extinct. On the western side of Yorkshire its habitats are confined to the narrow belt of magnesian limestone, on which it breeds regularly at Hackfall, near Ripon, and is reported as of rare occurrence at Marr, Conisborough, Watchley Crags, and Brockerdale; in the last-named localities it has been extinct, and the habitats restocked by specimens being turned down. As a straggler it has been found in various localities throughout the county, but many of these have been escaped or introduced specimens. Formerly it inhabited numerous localities in which it has now been for many years unknown.

Fam. URSIDÆ.

Ursus arctos *L.* **Brown Bear.**

There is no evidence whatever to show that the bear inhabited Yorkshire, beyond the fact that bones were found at Richmond, in a refuse-heap, which Prof. W. Boyd Dawkins considers to be of the date of the Roman occupation. The statements of the historians of Galtres Forest, that it was a famous harbour for bears, are open to very considerable doubt.

Sub-order *PINNIPEDIA*.

Fam. **TRICHECHIDÆ.**

29. Trichechus rosmarus *L.* Walrus.

Fam. **PHOCIDÆ.**

30. Phoca vitulina *L.* Common Seal.

Casual visitant, of uncommon occurrence along the coast and in the Humber. In the early years of the present century seals bred in great numbers at the mouth of the Tees, and in 1802, as appears from a document, a copy of which Mr. T. H. Nelson has sent me, they interfered to such an extent with the salmon fishery that determined measures were proposed for their extirpation. There is no evidence to show that the extermination was so effected, but it is hardly probable that they would long survive the rapid rise of the Cleveland iron trade and the shipping industries of Middlesborough, and in all likelihood the decade 1830 to 1840 would be that of the final extinction of the seal as a permanent resident in Yorkshire, though solitary individuals have been observed to within the last twenty years.

31. Phoca hispida *Schreb.* Ringed or Marbled Seal.

32. Phoca grœnlandica *Fab.* Greenland or Harp Seal.

33. Halichœrus gryphus (*Fab.*). Grey Seal.

In 1808 Graves, in his list of Cleveland animals, included not only the Common Seal but the 'Great Seal or Sea Calf' of Pennant's Zoology, 36.

Mr. R. M. Middleton, jun., informs me that in 1871 one was found alive at Seaton Snook, on the Durham shore of the Tees mouth.

34. Cystophora cristata (*Erxl.*). Hooded Seal.

Order CETACEA.

Sub-order *MYSTACOCETI.*

Fam. **BALÆNIDÆ.**

35. Balæna biscayensis *Eschricht.* Atlantic Right-whale.

Fam. **BALÆNOPTERIDÆ.**

36. Megaptera longimana (*Rudolphi*). Hump-backed Whale.

37. Balænoptera musculus (*L.*). Common Rorqual.

Casual visitant, of rare occurrence. A young one caught at Bridlington, April 5th, 1880, 16 feet long, is recorded by the name of *Balænoptera boops* Flem., or northern rorqual (E. Howarth, Nat., 1880, p. 26). Probably some of the whales described as 'rorquals,' 'finners,' &c., caught or seen, may be of this species, but specimens are seldom examined.

38. Balænoptera sibbaldii (*Gray*). Sibbald's Rorqual.

Casual visitant, of extremely rare occurrence. One, a young one, about 50 feet long, the skeleton of which is in the Hull Museum, was taken in the Humber, and described by Dr. Gray.

39. Balænoptera borealis *Less.* Rudolphi's Rorqual.

40. Balænoptera rostrata (*Fab.*). Lesser Rorqual.

Casual visitant, probably not unfrequent.

Dogger-bank, one 17 feet long, described by John Hunter, Phil. Trans., 1787. (Gray, Cat. of Seals and Whales, 2nd Ed., p. 193).

Hull, one, young, taken in the Victoria Dock Basin, Sep. 9, 1869, skeleton in Hull Museum.

Sewerby, near Bridlington, one, 16 or 17 feet long, stranded, spring of 1859 (Rev. Yarburgh Lloyd Greame, MS.).

Sub-order *ODONTOCETI.*

Fam. **PHYSETERIDÆ.**

Sub-fam. *PHYSETERINÆ.*

41. Physeter macrocephalus *L.* Sperm Whale.

Only one occurrence. An adult male, 56 feet long, was cast ashore at Tunstall, in Holderness, in 1825 ; the skeleton, 47 feet 7 inches, is preserved at Burton Constable.

42. **Hyperoödon rostratum** (*Chemnitz*). **Common Beaked-Whale.**

Casual visitant, of rare occurrence. A female was stranded in the Humber, close to Hull, in 1837, the skeleton of which is now in the Hull Museum (Gray, Cat. of Seals and Whales, 1866, p. 331 : cf. Zool., 1849, pp. 2409, 2441). Another specimen, stranded in Patrington Haven, was seen by Dr. Foster, by whom the skeleton was articulated. (Howarth, Nat., 1880, vol. vi., p. 26).

43. **Hyperoödon latifrons** *Gray.* **Broad-fronted Beaked-Whale.**

A skull, dredged up by a smack on the Great Fisher Bank, on March 15, 1881, from a depth of 36 fathoms, is now in the Norwich Museum (Southwell, Zool., June 1881, p. 258).

44. **Ziphius cavirostris** *Cuv.* **Cuvier's Whale.**

45. **Mesoplodon sowerbiensis** (*Blainville*). **Sowerby's Whale.**

<h3 align="center">Fam. DELPHINIDÆ.</h3>

Sub-fam. BELUGINÆ.

46. **Monodon monoceros** *L.* **Narwhal.**

Very doubtful. Mr. Thos. Waller, of Scarborough, informs me that one was stranded on the rocks at Flamborough in 1806, and that the horn passed into the possession of Mr. Arthur Strickland.

47. **Delphinapterus leucas** (*Pall.*). **White Whale.**

Sub-fam. DELPHININÆ.

48. **Orca gladiator** (*Lacép.*). **Killer. Grampus.**

Casual visitant, of frequent but irregular occurrence; ascends the Humber, even as high as Goole, in pursuit of salmon, and is often stranded.

49. **Grampus griseus** (*G. Cuv.*). **Risso's Grampus.**

50. **Globicephalus melas** (*Trail*). **Pilot Whale.**

Casual visitant, of rare occurrence. In June, 1862, a shoal of forty or fifty were stranded on the Whitton Sands, in the Humber, nearly opposite the mouth of the Trent; and a similar shoal went ashore at Cleethorpes, on the Lincoln-shire shore of the Humber mouth (G. Norman, Zool., 1862, p. 8087).

51. **Phocæna communis** *F. Cuv.* **Porpoise.**

Resident off the coast, very abundant. Great numbers follow the salmon up the Humber; sometimes ascends the Ouse and Wharfe as far as Cawood and Kirkby Wharfe.

52. **Delphinus delphis** *L.* **Common Dolphin.**

53. **Delphinus tursio** *Fab.* **Bottle-nosed Dolphin.**

Two caught at Spurn Point, in September, 1879, were seen by Mr. E. Howarth, and described by him in the Naturalist (Sept. 1880, vi. 26).

54. **Delphinus acutus** *Gray.* **White-sided Dolphin.**

55. **Delphinus albirostris** *Gray.* **White-beaked Dolphin.**

A young female, caught off Great Grimsby, in September, 1875, was figured and described by Dr. Cunningham (P.Z.S., 1876, pp. 679 to 686, and plate). The skeleton is in Edinburgh University Museum.

Order **UNGULATA.**

Sub-order *ARTIODACTYLA.*

Fam. **SUIDÆ.**

Sus scrofa *L.* **Wild Boar.**

Extinct, formerly frequented the forests of Yorkshire in large numbers, especially that of Galtres, in the vale of York. Its tusks are very frequently found there and in the alluvial deposits of Holderness. The direct testimony as to the existence of wild boars is weaker and less voluminous than in the case of the wolf, but the great mass of evidence of other kinds—etymological and legendary—suffices to warrant us in regarding it as formerly common in the county.

Fam. **CERVIDÆ.**

Rangifer tarandus *L.* **Reindeer.**

56. **Cervus elaphus** *L.* **Red Deer.**

Semi-domesticated, only in parks. Formerly wild, ranging over the whole county. Have been for centuries kept in Wharncliffe Chase and in Bolton Deer Park; and in these localities are, in all probability, the lineal descendants of the wild stock which formerly inhabited the surrounding districts. A deed of agreement between the Countess dowager of Pembroke and the Countess of Cork, dated May 20, 1654, shows the probable date at which the deer of Wharfedale were walled in at Barden and Bolton.

57. Cervus dama *L.* Fallow Deer.

Domesticated, only in parks.

58. Capreolus capræa *Gray.* Roe Deer.

Domesticated, in a few parks only. Formerly inhabited the county in a wild state. In the time of Edward III. they were numerous in the forest of Pickering, and in 1340 were the subject of a prosecution instituted by the Crown against Henry de Percy, lord of the adjacent manor of Semere.

Bones have occurred in the Raygill cave, but only in very few instances anywhere.

Fam. BOVIDÆ.

59. Bos taurus *L.* Wild White Cattle.

Extinct. Two herds existed in Yorkshire till within recent years—at Gisburn in Craven, and at Burton Constable in Holderness, but in a state of semi-domestication. The latter herd perished of distemper shortly before the close of the last century. That at Gisburn Park continued to exist till 1859, when the last bull was killed, the breed having so degenerated from constant inter-breeding that it was impossible to keep them longer.

Order RODENTIA.

Sub-order *SIMPLICIDENTATA.*

Section SCIUROMORPHA.

Fam. SCIURIDÆ.

60. Sciurus vulgaris *L.* Squirrel.

Generally distributed, and common in woods and plantations.

Fam. **CASTORIDÆ.**

Castor fiber *L.* European Beaber.

The only grounds we possess for surmising that the beaver ever inhabited Yorkshire are afforded by place-names. Beverley is supposed to have derived its name from this animal; while Beaverholes and Beaverdike in the Forest of Knaresborough, and Beevor Hall or Beverhole, near Barnsley, have their derivation also ascribed to this source. The places so named appear to have been suitable to the habits of this animal.

Section MYOMORPHA.

Fam. **MYOXIDÆ.**

61. Muscardinus avellanarius (*L.*). Dormouse.

Generally but very thinly distributed over the county; more abundant in densely wooded districts. No doubt it is much overlooked from its retiring habits.

Fam. **MURIDÆ.**

Sub-fam. *MURINÆ.*

62. Mus minutus *Pall.* Harvest Mouse.

Very irregularly and thinly distributed, and scarce.

63. Mus sylvaticus *L.* Long-tailed Field Mouse.

Generally distributed and abundant.

64. Mus musculus *L.* Common House Mouse.

Universally distributed and abundant wherever there are human habitations.

65. Mus rattus *L.* Black Rat.

Extremely local, appearing to occur only at Stockton-on-Tees, where it is not unfrequent in one or two old buildings. Is reported as having been taken in various other places scattered irregularly over the county, but in the rural districts it is probable that the black variety of the Water Vole has been mistaken for it; and those reported for the manufacturing and seaport towns are importations.

66. Mus decumanus *Pall.* **Brown Rat.**

Universally distributed about human habitations, and very abundant.

Sub-fam. ARVICOLINÆ.

67. Arvicola amphibia (*L.*). **Water Vole.**

Generally distributed, common. The Black variety occurs in a few localities, and there is reason to believe that some at least of the 'Black Rats' reported are of this variety of the Water Vole.

68. Arvicola agrestis *De Selys.* **Common Field Vole.**

Generally distributed, abundant.

69. Arvicola glareolus (*Schreb.*). **Red Field Vole.**

Reported from a few localities scattered irregularly over the whole county; probably more general, but not usually distinguished from the Common Field Vole.

Sub-order DUPLICIDENTATA.

Fam. LEPORIDÆ.

70. Lepus europæus *Pall.* **Common Hare.**

Generally distributed, common. Attains to a great size on the high wolds. White and pied varieties are not unfrequently reported.

71. Lepus variabilis *Pall.* **Varying Hare.**

72. Lepus cuniculus *L.* **Rabbit.**

Generally distributed, very abundant. Introduced on Penyghent and other places. Black and silver-grey varieties are of not unfrequent occurrence.

BIRDS.

WM. EAGLE CLARKE.

Class 2.   AVES.

Sub-class *AVES CARINATÆ.*

Series ÆGITHOGNATHÆ.

Order 1.   PASSERES.

Sub-order *OSCINES.*   •

Section 1.   OSCINES DENTIROSTRES.

Family **TURDIDÆ.**

Sub-family *TURDINÆ.*

1. **Turdus viscivorus** *L.*   **Missel-Thrush.**

Resident, generally distributed, abundant.   Its numbers are annually increased in early autumn by arrivals from the north, which on the advent of winter move further south.

2. **Turdus musicus** *L.*   **Song-Thrush.**

Resident, generally distributed, common; much less numerous in the winter, on the approach of which many migrate. Immigrants appear from the north annually on the coast, in the early autumn departing further south.

3. **Turdus iliacus** *L.*   **Redwing.**

Winter visitant, arriving usually in small parties in October and November, and departing in April.   Is recorded as having nested at Kildale in 1840 (Zool. 1845, p. 1056), at Glaisdale in 1872, when the nest, eggs, and parent bird were obtained (Zool. 1873, p. 3411), and supposedly near York in 1879 (Zool. 1879, p. 460).

4. **Turdus pilaris** *L.*   **Fieldfare.**

Winter visitant, usually arriving in flocks in October and November, often remaining in the spring until the first or second week of May.

c

5. **Turdus varius** *Pall.* **White's Thrush.**

Accidental visitant from Eastern Asia, of extremely rare occurrence.

Huddersfield, one (Beaumont, Hudd. Nat., 1864, p. 217).

Danby-in-Cleveland, one seen, spring of 1870 (Atkinson, Zool., May, 1870, p. 2142).

Whitby, one, November 1878 (Simpson, Zool., 1880, p. 68); now in the Whitby Museum.

6. **Turdus atrigularis** *Temm.* **Black-throated Thrush.**

7. **Turdus merula** *L.* **Blackbird.**

Resident, generally distributed, abundant. Immigrants from Northern Europe arrive annually on the coast in October and November, and sometimes later in the winter ; old males appearing later than the young ones.

8. **Turdus torquatus** *L.* **Ring-Ouzel.**

Summer visitant, nesting commonly on all the high moors of Cleveland, and the western moorlands from Sheffield northwards ; also occurring in limited numbers on the coast in late autumn, as an immigrant from the Continent, on its way further south. In the winter of 1855–6 single birds were observed at Holmfirth and at Keighley. Has nested in solitary instances on Thorne Moor and near Beverley, both low-lying localities, only a few feet above sea-level.

9. **Monticola saxatilis** *(L.).* **Rock-Thrush.**

Accidental visitant from Central and Southern Europe, of extremely rare occurrence.

Near Robin Hood's Bay, one, June 1852 (Bedlington, Morris' Nat., 1856, p. 21). Probably an adult male.

Sub-fam. *CINCLINÆ.*

10. **Cinclus aquaticus** *Bechst.* **Common Dipper.**

Resident, nesting commonly in the hilly districts of the north-east and of the west from Sheffield northwards. In extremely severe winters descends from the higher localities, and is then occasionally observed on the polluted streams of the manufacturing districts. Has never yet been reported as having occurred in the East Riding.

BIRDS.

BIRDS.

11. **Cinclus melanogaster** *C. L. Brehm.* **Black-bellied Dipper.**

Accidental visitant from Scandinavia, of rare occurrence; in Eastern Yorkshire only.

Welwick, one, Oct. 24, 1874, in the collection of Mr. P. W. Lawton, of Easington.

Beverley, one, Oct. 29, 1875 (Boyes, Zool. 1876, p. 4871).

Flotmanby, near Filey, Dec. 8, 1875 (Tuck, Field, Jan. 1876, p. 22).

Bridlington, one in the collection of Mr. J. H. Gurney, jun.

It is highly probable that the Dipper, recorded by the Rev. F. O. Morris (Nat. 1856, p. 186) as shot at Nunburnholme, was of this species, but the specimen not being now in existence, the question must remain an open one.

Sub-fam. *SAXICOLINÆ.*

12. **Saxicola œnanthe** (*L.*). **Common Wheatear.**

Summer visitant, local, being confined to uncultivated lands, but common where found. Arrives during March and April, departing in September and October, the young preceding the old birds in the autumn.

13. **Saxicola albicollis** *Vieill.* **Black-eared Chat.**

14. **Pratincola rubetra** (*L.*). **Whin-Chat.**

Summer visitant, generally distributed, common. Arrives in mid-April, and departs in October.

15. **Pratincola rubicola** (*L.*). **Stone-Chat.**

Resident, extremely local, and eccentric in its distribution. In the autumn the majority migrate, only a few remaining through the winter, and these more particularly near the coast. During the breeding season this species seems to affect waste lands at low or moderate elevations.

16. **Ruticilla phœnicurus** (*L.*). **Redstart.**

Summer visitant, generally but somewhat locally and thinly distributed. Arrives in April, and departs in August and September, the young birds leaving before the old ones.

17. **Ruticilla titys** (*Scop.*). **Black Redstart.**

Casual visitant, of uncommon occurrence on the coast, in spring, autumn, and winter. Mr. M. Bailey, of Flamborough, has frequently observed these birds in spring on their arrival on the headland, and has known them killed by flying against the light in thick, foggy weather, with the wind E.N.E. He has also seen them on their departure in September, and has noticed several in October and November. Inland, it is extremely rare, being recorded as observed near Leeds (several times, two or three in 1843), and once near Bingley.

Sub-fam. *SYLVIINÆ.*

18. **Cyanecula wolfi** *C. L. Brehm.* **White-spotted Bluethroat.**

Accidental visitant from Central and Western Europe, of extremely rare occurrence.

Near Scarborough, a female picked up dead beneath the telegraph wires, about the 9th of April, 1876 (Tuck, Zool., 1876, p. 4956).

19. **Cyanecula suecica (***L.***).** **Red-spotted Bluethroat.**

20. **Erithacus rubecula (***L.***).** **Redbreast.**

Resident, generally distributed, abundant. Immigrants are observed on the coast in the autumn, often in large numbers, returning early in March.

21. **Daulias luscinia (***L.***).** **Nightingale.**

Summer visitant, of regular occurrence in very limited numbers in the neighbourhood of Barnsley, Wakefield, York, Beverley, Patrington, Brough, Selby, and Doncaster, arriving early in May. West and north of the frontier formed by these towns it is only of exceptional occurrence, and a line drawn from Huddersfield, through Bradford, Otley, and Ripon, to Baldersby, Bagby, and Sessay, near Thirsk, and thence to Flamborough Head, will include all the localities for which there is satisfactory evidence of its ever having occurred or bred, and also defines the extreme northern limit of its distribution in Britain.

The Hon. F. H. Dawnay informs me that a pair passed the summer of 1868 at Baldersby Park, in 1876 it occurred at Sessay, and this year (1881) I am told by Mr. Robert Lee that it has appeared at Bagby—all near Thirsk. In 1876 it nested at Little Thorp, near Bridlington, an egg being sent for my inspection by Mr. W. F. Foster. There is also reliable evidence of its occurrence in other localities, not, however, quite so far north, and it seems well established that the nightingale is gradually extending its range up the vale of York and along the coast.

Early in the present century Doncaster was regarded by all writers as the most northerly locality which it visited annually, until in 1844 Mr. Thomas Allis showed that it occurred regularly much further north, and in two exceptional instances to as high a latitude as five miles north of York. Baldersby and Bagby, which must now be regarded as its northern outposts, are 21 miles N.W. of York, or 15 and 20 miles respectively north of the latitude of that city.

22. Sylvia rufa (*Bodd.*). Whitethroat.

Summer visitor, generally distributed, abundant. Arrives late in April, departing in August and September.

23. Sylvia curruca (*L.*). Lesser Whitethroat.

Summer visitor, generally distributed, though in varying numbers, and not nearly so abundant as *S. rufa*. Arrives during the last week of April, and departs in August and September.

24. Sylvia orphea *Temm.* Orphean Warbler.

Accidental visitor from Central and Southern Europe and Northern Africa, of extremely rare occurrence.

Wetherby, male seen, female shot, July 6, 1848 (Milner, Zool., 1849, p. 2588).

Notton Wood, near Wakefield, a nest with four eggs taken, June, 1864 (Harting, Field, April 22, 1871, p. 321).

25. Sylvia atricapilla (*L.*). Blackcap.

Summer visitor, somewhat irregularly distributed over the county, but not very numerous. Has in several instances occurred in mid-winter. Arrives late in April.

26. Sylvia salicaria (*L.*). Garden Warbler.

Summer visitor, generally distributed, common. Arrives early in May, leaving in September.

27. Sylvia nisoria *Bechst.* Barred Warbler.

28. Melizophilus undatus (*Bodd.*). Dartford Warbler.

Casual visitor, observed in one locality only—the Rivelin Valley—in the extreme south. Here Mr. Charles Dixon, who is well acquainted with the bird, has several times seen

it in the gorse coverts, and in one solitary instance he found
a nest with five eggs, observing the sitting bird from a
distance of only a few feet.   Hitherto Melbourne in Derbyshire has been considered the most northerly locality in
which it has occurred.

Sub-fam. *PHYLLOSCOPINÆ*.

29. Regulus cristatus *Koch.*   Golden-crested Wren.

Resident, generally distributed in suitable localities, but not
very numerous ; is also a winter visitant, arriving regularly
on the coast about the middle of October, sometimes in
immense flocks.

30. Regulus ignicapillus (*C. L. Brehm*).   Fire-crested Wren.

Casual visitant, of extremely rare occurrence in the winter.

Whixley, one, Dec. 3rd, 1849 (Garth, Zool., 1849, p. 2699).

Huddersfield, one said to have occurred at Armitage Bridge,
Sept. 3rd, 1874 (Varley, Nat., 1875, p. 24).

Whitby, one in the local collection at the Museum
(Stephenson, MS.).

Is also said to have occurred at Woodend, near Thirsk (Allis,
1844).   There are other records, but the species is much
confounded with old males of *R. cristatus.*

31. Phylloscopus superciliosus (*Gm.*).   Yellow-browed Warbler.

32. Phylloscopus collybita (*Vieill.*).   Chiffchaff.

Summer visitant, generally distributed, and common throughout the woodland districts of central and eastern Yorkshire, ranging there from north to south; much less
numerous in the south-west, while in the north-western
dales it is of very rare and exceptional occurrence.

33. Phylloscopus trochilus (*L.*).   Willow-Wren.

Summer visitant, generally distributed, very abundant.
Arrives in the middle of April, and leaves late in September or early in October.

34. Phylloscopus sibilatrix (*Bechst.*).   Wood-Wren.

Summer visitant, local, but not uncommon in woodland districts where the trees are of considerable growth.   Occurs
as high as 1350 feet in the woods above Malham Tarn.
Arrives after the middle of April.

Sub-fam. *ACROCEPHALINÆ.*

35. Hypolais icterina (*Vieill.*). Icterine Warbler.

36. Aedon galactodes (*Temm.*). Rufous Warbler.

37. Acrocephalus streperus (*Vieill.*). Reed-Warbler.
   Summer visitant, very locally distributed; numerous where it occurs. Breeds regularly at Hornsea Mere; also near Hull and other localities in Holderness; at Swinefleet; near Leeds; at Knaresborough; and formerly at Scarborough Mere. In numerous other localities in the county it has bred occasionally.

38. Acrocephalus palustris (*Bechst.*). Marsh-Warbler.

39. Acrocephalus arundinaceus (*L.*). Great Reed-Warbler.

40. Acrocephalus aquaticus (*Gm.*). Aquatic Warbler.

41. Acrocephalus schœnobænus (*L.*). Sedge-Warbler.
   Summer visitant, generally distributed, very common. Arrives in the beginning of May, departs at the end of September or early in October.

42. Locustella nævia (*Bodd.*). Grasshopper Warbler.
   Summer visitant, locally and thinly distributed, and in some localities uncertain in its appearance.

43. Locustella luscinioides (*Savi*). Savi's Warbler.

Fam. **ACCENTORIDÆ.**

44. Accentor collaris (*Scop.*). Alpine Accentor.
   Accidental visitant from Central and Southern Europe, of extremely rare occurrence.
   Scarborough, female, winter of 1862-3 (Boulton, Zool, 1863, p. 8766).

45. Accentor modularis (*L.*). Hedge-Sparrow.
   Resident, generally distributed, common; a regular autumn immigrant on the Holderness coast.

Fam. **PANURIDÆ.**

46. Panurus biarmicus (*L.*). Bearded Reedling.
   Casual visitant, of extremely rare occurrence.

Walton Hall, Wakefield, a pair once bred by the side of the lake (More, *fide* Waterton, Ibis, 1865).

Sheffield, a pair seen, Dec., 1878 (Dixon, MS.).

Also reported to have occurred at Scarthingwell (Chaloner, MS.), and at Kirkleatham (Zool., 1845, p. 1135).

Fam. PARIDÆ.

47. Acredula rosea (*Blyth*).   Long-tailed Titmouse.

Resident, generally distributed, fairly common.   Most frequently seen in autumn and winter.

48. Acredula caudata (*L.*).   Continental Long-tailed Titmouse.

49. Parus major *L.*   Great Titmouse.

Resident, generally distributed, common.   Sometimes there are considerable arrivals of immigrants on the coast in autumn.

50. Parus ater *L.*   European Coal Titmouse.

51. Parus britannicus *Sharpe and Dresser*.   English Coal Titmouse.

Resident, common, generally distributed; but in some localities in south-west Yorkshire it is of rare occurrence in summer, being most frequently observed as a winter visitant.   In Holderness, additions to its numbers by immigration are observed in the autumn.

52. Parus palustris *L.*   Marsh-Titmouse.

Resident, generally distributed, but in varying abundance.

53. Parus cæruleus *L.*   Blue Titmouse.

Resident, generally distributed, abundant; also observed as an autumn immigrant.

54. Lophophanes cristatus (*L.*).   Crested Titmouse.

Casual visitant, of very rare occurrence.

Yorkshire, said to have occurred in the county by Lewin (British Birds, vol. 5, p. 46).

Thorne, one seen (Allis, 1844).

Thornton Moor, near Bradford, March, 1870 (Butterfield, MS.).

Whitby, one, March, 1872 (Simpson, Zool., 1872, p. 3021).

Mr. Thomas Stephenson states that the specimen in the local collection at the Whitby Museum was obtained on the Newton House estate, which abounds with larch plantations; and he has it on the authority of Parker, formerly keeper there, that the bird was never lost sight of either in winter or summer, and he (Parker) had no doubt they bred there.

Thirsk, one shot, preserved by Mr. Robert Lee (Lee, MS.).

Fam. SITTIDÆ.

55. Sitta cæsia *Wolf.* Common Nuthatch.

Resident, local, and far from numerous. Chiefly confined to the older parks, such as Castle Howard, Hovingham, Swinton, Hackfall, Walton, Stainborough, and Wharncliffe. Entirely absent from the East Riding.

Fam. CERTHIIDÆ.

56. Certhia familiaris *L.* Creeper.

Resident, generally distributed in wooded districts, but far from numerous.

Fam. TROGLODYTIDÆ.

57. Troglodytes parvulus *Koch.* Wren.

Resident, generally distributed, common; annually observed on the coast as an autumn immigrant.

Fam. MOTACILLIDÆ.

58. Motacilla alba *L.* White Wagtail.

Casual visitant, of rare occurrence in summer; probably much overlooked, the following being all the instances in which it has been noted:—York, one, July 13, 1848 (Webb, Zool., 1848, p. 2229); Wakefield, several occurrences, has once nested there (Talbot, Birds of Wakefield, 1876); Bolton Abbey, one, April 12, 1879 (Clarke, Zool., 1880, p. 355); Gisburn, one, April 18, 1881.

59. Motacilla lugubris *Temm.* Pied Wagtail.

Resident, generally distributed, abundant in summer, comparatively few in the winter, for in the autumn the majority depart south, returning in early spring.

60. **Motacilla melanope** *Pall.*   **Grey Wagtail.**

Resident, generally distributed in the winter, in summer confined to the high lands of the west, from Sheffield northwards, and of Cleveland.

61. **Motacilla flava** *L.*   **Blue-headed Wagtail.**

62. **Motacilla raii** *Bonap.*   **Yellow Wagtail.**

Summer visitant, fairly general in its distribution, but in varying numbers, most abundant in agricultural and pastoral districts ; in some few localities only observed occasionally.   Arrives in mid-April, departing in considerable flocks in September.

63. **Anthus pratensis** (*L.*).   **Meadow-Pipit.**

Resident, generally distributed, abundant.   Considerable flocks move southwards in September, returning early in March; those observed in the winter, when it is much less numerous, are probably immigrants from more northern districts.

64. **Anthus trivialis** (*L.*).   **Tree-Pipit.**

Summer visitant, generally distributed, numerous in wooded districts.   Arrives during the second week of April, leaves in small flocks in September.

65. **Anthus campestris** (*L.*).   **Tawny Pipit.**

Accidental visitant from Continental Europe and Northern Africa, of extremely rare occurrence.

Near Bridlington, a male shot, Nov. 20, 1869, by Mr. Thos. Boynton, and now in his possession (Zool., 1870, pp. 2021, 2068, 2101).

66. **Anthus richardi** *Vieill.*   **Richard's Pipit.**

Casual visitant, of extremely rare occurrence.   One is said, on the authority of the late David Graham, to have occurred on the coast, in 1849 (Zool. 1849, p. 2569).

67. **Anthus spipoletta** (*L.*).   **Water-Pipit.**

68. **Anthus obscurus** (*Lath.*).   **Rock-Pipit.**

Resident on the coast, breeding abundantly at Flamborough, and no doubt also near Whitby, where it is observed all the year round ; generally distributed on the coast in autumn and winter.

Fam. PYCNONOTIDÆ.

69. **Pycnonotus capensis** (*L.*). **Gold-vented Bulbul.**

Fam. ORIOLIDÆ.

70. **Oriolus galbula** *L.* **Golden Oriole.**

Casual visitor, of very rare occurrence as a straggler during the spring and autumn migrations.

Spurn, female, spring of 1834 (Allis, 1844); now in York Museum.

Wakefield, one shot at Bottom Boat, beginning of August, 1856 (Talbot, MS.).

Hunmanby, adult male shot, May, 1859 (Roberts, Zool., 1859, p. 6561).

Doncaster, male seen April 28, 1870, by Rev. J. W. Chaloner (MS.).

Grimston Park, Tadcaster, one seen by the keeper, April, 1870 (Chaloner, MS.).

Swinton Park, Masham, one seen about May, 1870 (Carter, MS.).

Bingley, one observed, early autumn, 1875 (Butterfield, MS.).

Specimens are also said to have been obtained near Hull, Scarborough, and other localities on the coast.

Fam. LANIIDÆ.

71. **Lanius excubitor** *L.* **Great Grey Shrike.**

Winter visitor, occurring annually, in limited numbers, and most numerous in the vicinity of the coast on its arrival late in October, a few remaining through the winter, leaving in the early spring.

72. **Lanius minor** *Gmel.* **Lesser Grey Shrike.**

73. **Lanius collurio** *L.* **Red-backed Shrike.**

Casual visitor, of rare occurrence during the spring and autumn migrations. Has nested near Barnsley (1826), Richmond, Halifax, Huddersfield, Sheffield (1868), Silsden, and Beverley (1876). It appears to be much less frequent than formerly, when to certain localities it was regarded as almost an annual visitant.

74. **Lanius auriculatus** *Müll.* **Woodchat Shrike.**

Accidental visitor from Central and Southern Europe and Africa, of extremely rare occurrence.

Yorkshire, mentioned as having occurred (Yarrell, British Birds, 1843, *fide* Leadbeater).

Scarborough, two young birds obtained by Mr. A. S. Bell, 1860 or 1861 (Clarke, Birds of Yorkshire, p. 61).

Fam. AMPELIDÆ.

75. **Ampelis garrulus** *L.* **Waxwing.**

Casual visitor, appearing at irregular intervals during the winter months, entirely absent most years, while others are noticeable for their irruption in great flights. First recorded as British from a specimen obtained at York in Jan. 1681, by the celebrated Dr. Martin Lister, F.R.S. The years 1686, 1787, 1849–50, 1863–4, and 1866–7 were remarkable for the immense numbers in which it visited the county.

Fam. MUSCICAPIDÆ.

76. **Muscicapa grisola** *L.* **Spotted Flycatcher.**

Summer visitor, generally distributed, common. Arrives in May, leaves in September.

77. **Muscicapa atricapilla** *L.* **Pied Flycatcher.**

Summer visitor, very local in its distribution, fairly numerous where it occurs. Breeds annually at Castle Howard, Duncombe Park, Hovingham, Masham, Hackfall, Bolton Woods, in woods near Pateley Bridge, Stainborough and Cannon Hall Parks, near Barnsley, and occasionally in other localities in the West and North Ridings. In the East Riding it is observed only during migration in spring and autumn.

78. **Muscicapa parva** *Bechst.* **Red-breasted Flycatcher.**

Section 2. OSCINES LATIROSTRES.

Fam. **HIRUNDINIDÆ.**

79. Hirundo rustica *L.* Swallow.

Summer visitant, generally distributed, abundant. Arrives in the middle of April, leaving at the end of September, stragglers remaining till late in October.

80. Chelidon urbica (*L.*). Martin.

Summer visitant, generally distributed, abundant. Arrives late in April, leaves in September, stragglers late in October. Breeds under the cliff-ledges at Malham Cove and at Flamborough.

81. Cotile riparia (*L.*). Sand-Martin.

Summer visitant, abundant in suitable localities throughout the county. Has once bred in haystacks at Knapton Hall, near Malton.

82. Progne purpurea (*L.*). Purple Martin.

Accidental visitant from North America, of extremely rare occurrence.

Huddersfield, one shot at Colne Bridge, 1854 (Hobkirk's Hudd., 1859, p. 144). Requires investigation.

Section 3. OSCINES CONIROSTRES.

Fam. **FRINGILLIDÆ.**

Sub-fam. FRINGILLINÆ.

83. Carduelis elegans *Steph.* Goldfinch.

Resident, very local, far from numerous, much less abundant than formerly, having now disappeared from many of its old haunts. In some localities observed as an occasional autumn and winter visitant.

84. Chrysomitris spinus (*L.*). Siskin.

Winter visitant, rather uncertain in its appearance, and varying in its numbers. Has nested at Walton Hall, near Wakefield (More, *fide* Waterton, Ibis, 1865).

85. Serinus hortulanus *Koch.* Serin Finch.

86. Ligurinus chloris (*L.*). Greenfinch.

> Resident, generally distributed, abundant. Large arrivals of young and females are observed on the coast in autumn, returning late in April and May in considerable flocks, which then contain many old males.

87. Coccothraustes vulgaris *Pall.* Hawfinch.

> Resident, local, breeding regularly in many localities; more generally distributed in winter, when large flocks have occasionally been observed. Though not a numerous species, it is much more so than it was a few years ago, and now nests in some numbers in the neighbourhood of Beverley.

88. Passer domesticus (*L.*). Common Sparrow.

> Resident, generally distributed, extremely abundant.

89. Passer montanus (*L.*). Tree Sparrow.

> Resident, local, but not uncommon in Eastern and Central Yorkshire; rarer in the west, where it appears to be unknown in many districts. Commoner in the autumn and winter, immense flocks occasionally arriving on the coast from the north in October.

90. Fringilla cœlebs *L.* Chaffinch.

> Resident, generally distributed, abundant; partially migratory, flocks consisting of females and young arriving on the coast in autumn.

91. Fringilla montifringilla *L.* Brambling.

> Winter visitant, occurring annually in varying numbers, but in some inland districts is only occasionally observed. Females and young of the year arrive first, old males later in separate flocks. Is *said* to have nested in Baldersby Park, near Thirsk, in 1864 (Atkinson, Zool., 1864, p. 9210).

92. Linota cannabina (*L.*). Linnet.

> Resident, generally distributed, abundant in suitable localities. Immense flocks are observed near the coast in autumn, on their passage south.

93. Linota linaria (*L.*). Mealy Redpoll.

Winter visitant, of extremely irregular occurrence. Sometimes appears in large flocks, as in 1855, 1861, 1876.

94. Linota rufescens (*Vieill.*). Lesser Redpoll.

Resident, generally distributed and common; immigrants, probably from more northern British localities, arrive in autumn, at which season and throughout the winter it is usually observed in flocks.

95. Linota hornemanni *Holb.* Greenland Redpoll.

96. Linota flavirostris (*L.*). Twite.

Resident, sparingly scattered over the southern, western, and northern moorlands (including those of Cleveland) during the breeding season ; much more generally distributed in autumn and winter. Said by Mr. Allis (1844) to breed on Thorne Moor, a low-lying heath on the borders of Lincolnshire. In the autumn it occurs at Spurn, often in large flocks. In very severe winters moves further south, returning in the middle of March.

Sub-fam. LOXIINÆ.

97. Carpodacus erythrinus (*Pall.*). Scarlet Grosbeak.

98. Pyrrhula europæa *Vieill.* Bullfinch.

Resident, somewhat local, and not abundant ; observed to be more general in autumn and winter. Immigrants occasionally arrive in autumn, large numbers having been observed on the coast in 1880.

99. Pinicola enucleator (*L.*). Pine-Grosbeak.

Accidental visitant from Northern Europe, Asia, and America, of extremely rare occurrence.

Doncaster and Sheffield : In the sale catalogue (Dec. 28, 1866) of Mr. Sealy, of Cambridge, lot 59 is described as 'Pine Grosbeaks, three in a case, one shot at Doncaster and one at Sheffield' (J. H. Gurney, jun., Zool., 1877, p. 249).

Near Whitby, one shot by Mr. Kitching, about 1861, in the winter, now in the local collection at the Whitby Museum (Stephenson, MS.).

100. **Loxia pityopsittacus** *Bechst.* Parrot Crossbill.

Accidental visitant from Northern Europe, of rare occurrence.

Flamborough, female shot by Mr. Bailey, Aug. 4, 1866 (Boulton, Zool., 1867, p. 543).

101. **Loxia curvirostra** *L.* Common Crossbill.

Winter visitant, irregular both in appearance and numbers, sometimes appearing in immense flights. Occasionally solitary pairs remain to breed. Has nested at Boynton near Bridlington in 1829, and at Bramham, where several nests were found in 1840 (Allis). At Plumpton, near Harrogate, in July, 1876, I saw two old birds accompanied by four young ones. Young birds just from the nest have been observed by the keeper at Gilling-in-Ryedale (Brigham, MS.).

102. **Loxia leucoptera** *Gmel.* White-winged Crossbill.

Accidental visitant from North America, of rare occurrence.

Cowick, near Snaith, a flock seen, from which two males and two females were obtained, Dec. 27, 1845 (Milner, Zool., 1847, p. 1694).

103. **Loxia bifasciata** (*C. L. Brehm*). Two-barred Cross-bill.

Sub-fam. *EMBERIZINÆ.*

104. **Emberiza melanocephala** *Scop.* Black-headed Bunting.

105. **Emberiza miliaria** *L.* Common Bunting.

Resident, rather locally distributed, but common. Considerable arrivals occur in the autumn. I have observed that the species is most abundant in the immediate vicinity of the coast.

106. **Emberiza citrinella** *L.* Yellow Bunting.

Resident, generally distributed, abundant. Large arrivals are observed in autumn, migrating north again in spring, often in very large flocks.

107. **Emberiza cirlus** *L.* Cirl Bunting.

Casual visitant, of rare occurrence.

Campsall, Doncaster, female, April 25, 1837 (Allis).

Near York, one, Dec., 1840 (Allis).

Richmond, one at St. Agatha's Abbey, Feb., 1840 (Strang-wayes, Zool., 1851, p. 3056).

Askew Moor, Bedale, two males, Dec. 29, 1850 (Id.).

The Leases, Bedale, male, Feb. 5, 1851 (Id.).

Woodsome, near Huddersfield, a pair said to have nested in 1856 (Hobkirk, Hudd. Cat., 1859).

Bolton-on-Dearne, one shot, another seen, Jan. 12, 1881 (T. Lister, Nat., 1881, vi. 124).

108. **Emberiza hortulana** *L.* **Ortolan Bunting.**

Casual visitant, of extremely rare occurrence.

Guisborough Moors, Cleveland, a female, or young bird of the year, seen Aug. 16, 1863 (Atkinson, Zool., 1863, p. 8768).

'The specimen which served for Mr. Bewick's figure was caught at sea on the Yorkshire Coast' (Jardine, B. Birds, ii. 311).

109. **Emberiza rustica** *Pall.* **Rustic Bunting.**

110. **Emberiza pusilla** *Pall.* **Little Bunting.**

111. **Emberiza schœniclus** *L.* **Reed-Bunting.**

Resident, generally distributed, but not very numerous, and less so in winter. Migrates in autumn, being replaced by arrivals from the north; returning again early in April.

112. **Plectrophanes lapponicus** (*L.*). **Lapland Bunting.**

Accidental visitant from Northern Europe, Asia, and America, of rare occurrence.

Whitby, one shot in the spring, about 1870, now in the Whitby Museum (Stephenson, MS.).

113. **Plectrophanes nivalis** (*L.*). **Snow-Bunting.**

Winter visitant to the coast, in flocks mainly composed of females and young of the year, with comparatively few old males; arriving during the latter part of October. Inland its appearance is much more irregular, and most frequent in severe seasons.

Section 4.　OSCINES SCUTELLIPLANTARES.

Fam. ALAUDIDÆ.

114. Galerita cristata (*L.*).　Crested Lark.

115. Alauda arvensis *L.*　Sky-Lark.

Resident, generally distributed, abundant. Immense flocks of immigrants from the continent arrive on the coast in autumn, departing further south on the advent of winter.

116. Alauda arborea *L.*　Wood-Lark.

Resident, extremely limited both in numbers and distribution. Breeds at Brandsby, Duncombe Park, and Hackness near Scarborough, and occasionally at Maltby and Doncaster. Has occurred—chiefly in the winter—near Whitby, Bridlington, Leeds, Wakefield, Barnsley, and Halifax.

117. Calandrella brachydactyla (*Leisl.*).　Short-toed Lark.

118. Melanocorypha sibirica (*Gm.*).　White-winged Lark.

119. Otocorys alpestris (*L.*).　Shore-Lark.

Winter visitant, entirely confined to the coast, and irregular both in appearance and numbers. Abundant in the winter of 1879-80, when they arrived on Dec. 22nd and departed about the 20th of March.

Section 5.　OSCINES CULTRIROSTRES.

Fam. STURNIDÆ.

120. Agelæus phœniceus (*L.*).　Red-winged Starling.

Accidental visitant from the American continent, of extremely rare occurrence.

Adwick-le-Street, a male found, probably killed by the telegraph wire, March 31st, 1877 (S. L. Mosley, Zool., 1877, p. 257; Nat., 1877, p. 53).

121. **Sturnus vulgaris** *L.* **Common Starling.**

Resident, generally distributed, abundant. Immense flocks arrive in autumn—young in July and August, old in September and October; departing northward in spring.

122. **Pastor roseus** (*L.*). **Rose-coloured Starling.**

Casual visitant, of uncommon occurrence in autumn. Has been observed many times in East Yorkshire and near the coast, chiefly in August ; less frequently inland. Occurrences are too numerous to mention.

Fam. **CORVIDÆ.**

123. **Pyrrhocorax graculus** (*L.*). **Red-billed Chough.**

Casual visitant, of extremely rare occurrence.

Hatfield, one killed by Mr. R. Glossop's keeper, and stuffed by Hugh Reid (Allis, 1844).
Sheffield : 'has, to my knowledge, been once observed' (C. Dixon, MS.).

124. **Nucifraga caryocatactes** (*L.*). **Nutcracker.**

Accidental visitant from Northern and Central Europe, of extremely rare occurrence.

Campsall, one said to have occurred, on the authority of Mr. Neville Wood (Lankester's Askern, 1842, p. 70).
Wakefield, one, autumn of 1865, in Mr. Harting's collection, (Harting, Handbook of B. Birds, p. 119).

125. **Garrulus glandarius** (*L.*). **Common Jay.**

Resident, not uncommon in wooded districts ; occasionally observed as an autumn immigrant. In the manufacturing districts this species is fast decreasing in numbers, and extremely local.

126. **Pica rustica** (*Scop.*). **Magpie.**

Resident, generally distributed, and fairly abundant in spite of much persecution.

127. **Corvus monedula** *L.* **Jackdaw.**

Resident, generally distributed, common. Immigrants often arrive with the rooks in the autumn, but never in separate flocks. Breeds in immense numbers in the coast cliffs.

128. **Corvus corone** *L.* **Carrion-Crow.**

Resident, generally but thinly distributed. Local and scarce
in the manufacturing districts. A few breed in the cliffs of
the coast.

129. **Corvus cornix** *L.* **Hooded Crow.**

Winter visitant, most abundant on and near the coast, where
it arrives in October and November, leaving in March and
April. In many inland districts it is only of occasional
occurrence, whilst to others it is an annual visitor. Has
occasionally remained to breed, there being authentic
evidence of its having done so on the Hornby estate, near
Catterick (Nat., 1865, p. 101). At Hackness, near Scarborough, a female paired for three successive seasons with
a male *C. corone*—the young resembling one or the other
parent (W. C. Williamson, P.Z.S., 1836, p. 76). There are
also several instances of its breeding at Flamborough, and
possibly in other localities where it has been observed
throughout the summer.

130. **Corvus frugilegus** *L.* **Rook.**

Resident, generally distributed, abundant. Immigrants in
large numbers come in from the continent in October and
November. In 1846 two pairs bred on chimneys in the
town of Hull.

131. **Corvus corax** *L.* **Raven.**

Resident, now restricted to a very few pairs in the North-
Western Fells, and possibly a pair may also still exist in
Cleveland. At the commencement of the present century
it was fairly general in its distribution, breeding in woods,
fells, and sea-cliffs, and even on the mausoleum at Castle
Howard. Owing to the persecution to which this species
has been subjected, its extermination as a Yorkshire bird
is now only a question of a few years.

Order 2. **MACROCHIRES.**

Fam. **CYPSELIDÆ.**

132. **Cypselus apus** (*L.*). **Common Swift.**

Summer visitant, generally distributed and common, except
in the manufacturing districts, where it is more or less
scarce. Arrives in the first week of May, occasionally
remaining as late as the second week of October. A few
nest in the cliffs at Flamborough.

133. **Cypselus melba** (*L.*). **White-bellied Swift.**

Accidental visitant from Central and Southern Europe, of rare occurrence.

Ripponden, near Halifax, one caught by the late Mr. Priestley, in the autumn of 1872, is now in his widow's possession (Rawson, MS.).

Scarborough, one, first seen on the 17th of April, 1880, remaining for a fortnight (West, Zool., 1880, p. 407). In a letter to me, Mr. West so accurately describes the bird as to leave no doubt as to its identification.

Hornsea Mere, one seen by Mr. F. Boyes.

134. **Acanthyllis caudacuta** (*Lath.*). **Needle-tailed Swift.**

Fam. **CAPRIMULGIDÆ.**

135. **Caprimulgus europæus** *L.* **Common Nightjar.**

Summer visitant, decidedly local in its distribution, and not numerous, affecting the woodland, moorland, and fell districts.

136. **Caprimulgus ruficollis** *Temm.* **Russet-necked Night-jar.**

Order 3. **PICI.**

Fam. **PICIDÆ.**

Sub-fam. *PICINÆ.*

137. **Dryocopus martius** (*L.*). **Great Black Woodpecker.**

Doubtful; if it occurs at all is an accidental visitant from Continental Europe, of extremely rare occurrence. Is said to have occurred in the following instances:—

Yorkshire, once in (Fothergill, Ornith. Brit., 1799, p. 3).

Yorkshire, one shot (Yarrell, 1843, ii. 128).

Yarm, two seen (Hogg, Zool., 1845, p. 1107).

Ripley, one killed, March 1846 (Garth, Zool., 1846, p. 1298).

138. **Picus major** *L.* **Great Spotted Woodpecker.**

Resident, local, thinly though more generally distributed
than the other species in Yorkshire; also observed on the
coast as an autumn immigrant.

139. **Picus minor** *L.* **Lesser Spotted Woodpecker.**

Resident, extremely local, confined to deeply wooded
localities, in which it appears to occur in very limited
numbers. Seems to be entirely absent from the East
Riding, and also from the valleys of the north-west; occurs
as far north as Thirsk and Slingsby.

Picus villosus *L.* **Hairy Woodpecker.**

Doubtful, a native of Eastern North America.

Kirklees Hall, near Brighouse, a pair shot, which passed into
the collection of the Duchess of Portland (Latham, Gen.
Syn., ii. 578).

Whitby, one early in 1849 (Higgins, Zool., 1849, p. 2496;
Bird, tom. cit., 2527; Newman, op. cit., 1851, p. 2985;
Bird, tom. cit., p. 3034).

140. **Gecinus viridis** (*L.*). **Green Woodpecker.**

Resident, local, but fairly numerous where it occurs.

Sub-fam. *IYNGINÆ*.

141. **Iynx torquilla** *L.* **Wryneck.**

Summer visitant, extremely local, being confined in the
breeding season to the south-eastern portion of the West
Riding and the adjacent portion of the East Riding,
where it is very sparsely distributed. During the spring
and autumn migrations it is occasionally observed on and
near the coast. Has been shot at Danby-in-Cleveland
during the breeding season. Appears to be now much
less frequent than formerly; Mr. H. Denny, in 1840,
describing it as formerly tolerably frequent in the neighbourhood of Leeds; and Mr. John Hogg, in 1845, wrote
of it as not uncommon in north-west Cleveland and south-
east Durham.

Series DESMOGNATHÆ.

Order 1. COCCYGES.

Sub-order *COCCYGES ANISODACTYLI.*

Fam. ALCEDINIDÆ.

142. **Alcedo ispida** *L.* **Common Kingfisher.**
Resident, generally though sparingly distributed. Almost banished from its former haunts in the manufacturing districts by the pollution of streams and by persecution; here, however, it is observed as a straggler in winter. Early in August and during September it often appears in some numbers on the Holderness drains.

Fam. CORACIIDÆ.

143. **Coracias garrulus** *L.* **Common Roller.**
Casual visitant, of uncommon occurrence in summer. The localities for which it has been recorded are :—Fixby, in the winter of 1824; Seamer, 1832; Scarborough, June, 1833; off the coast, May, 1843; Hatfield (Allis, 1844); Skelton Castle, near Redcar, a pair, July, 1847; Whitby, 1852; Bridlington, 1868; Bingley, 1872; Grosmont, June, 1874: Marske Hall, Richmond; and Thirsk, June 5, 1880.

Fam. MEROPIDÆ.

144. **Merops apiaster** *L.* **Common Bee-eater.**
Accidental visitant from Southern Europe and Northern Africa, of extremely rare occurrence.

Sheffield, one, about 1849 (Morris, B. Birds, i. 313).

Filey, a male picked up exhausted, June 9, 1880 (R. Richardson, MS.).

Fam. UPUPIDÆ.

145. **Upupa epops** *L.* **Hoopoe.**
Casual visitant, of uncommon occurrence as a straggler during the spring and autumn migrations, principally on the coast, but now less frequently than formerly. The occurrences are too numerous to mention.

Sub-order *COCCYGES ZYGODACTYLI.*

Fam. CUCULIDÆ.

146. **Cuculus canorus** *L.* **Cuckoo.**

Summer visitant, generally distributed, common. Arrives
in the third week of April, departing early in August.
Young birds linger to the end of September, and even into
October.

147. **Coccystes glandarius** (*L.*). **Great Spotted Cuckoo.**

148. **Coccyzus americanus** (*L.*). **Yellow-billed Cuckoo.**

149. **Coccyzus erythrophthalmus** (*Wils.*). **Black-billed
Cuckoo.**

Order 2. ACCIPITRES.

Sub-order *STRIGES.*

Fam. STRIGIDÆ.

150. **Strix flammea** *L.* **Barn-Owl.**

Resident, generally distributed, fairly common; most
numerous in South Yorkshire. In the latter months of the
year there are instances of this species sometimes occurring
in unusual numbers inland, and it is also occasionally
observed on the coast as an autumn immigrant.

Fam. BUBONIDÆ.

151. **Asio otus** (*L.*). **Long-eared Owl.**

Resident in woodland districts, where it is local in its distribution but common where found. Is observed annually
on the coast as an immigrant in late autumn, accompany-
ing *A. accipitrinus,* and in the winter; and it is questionable if these immigrants do not replace the local birds,
which appear to leave in the winter and return in early
spring.

152. **Asio accipitrinus** (*Pall.*). **Short-eared Owl.**

Winter visitant, generally distributed, common, arrives from the north in October and November. Occasionally it remains to breed, having nested in several localities in Cleveland, near Scarborough, on Thorne Waste, and once near Otley.

153. **Syrnium aluco** (*L.*). **Tawny Owl.**

Resident, generally distributed, and fairly numerous in well-wooded localities, excepting those of the manufacturing districts, where it is local. The commonest Yorkshire owl.

154. **Nyctea scandiaca** (*L.*). **Snowy Owl.**

Accidental visitant from Northern Europe and America, of extremely rare occurrence.

Barlow Moor, near Selby, one shot, Feb. 13, 1837, and exhibited at the Zoological Society in the same year (Clapham, MS.). Mr. Clapham tells me that the statement in Morris' British Birds (i. 195)—that this bird was accompanied by another of the same species—is incorrect. The specimen is now in the Leeds Museum.

There is evidence to show that this species has probably occurred on three other occasions in the county—at Scarborough and Filey about thirty years ago, at Flamborough in October, 1865, and again at Scarborough in Dec., 1879 (See Birds of Yorkshire, pp. 55, 56).

155. **Surnia ulula** *L.* **Hawk Owl.**

156. **Surnia funerea** *L.* **American Hawk Owl.**

157. **Nyctala tengmalmi** (*Gm.*). **Tengmalm's Owl.**

Accidental visitant from Northern Europe, of extremely rare occurrence.

Sleights, near Whitby, one about 1840, formerly in the Whitby Museum (Stephenson, MS.).

Hunmanby, one shot about 1847 (B. R. Morris, Zool., 1849, p. 2649).

Flamborough, one caught October 1, 1863 (Boulton, Zool., 1864, p. 9020); in the collection of Mr. John Stevenson, of Beverley.

Egton, near Whitby, one shot Nov. 19, 1872 is in the collection of Mr. W. Lister, of Glaisdale (Birds of Yorksh., p. 44).

Normanby, near Whitby, one shot Dec. 30, 1880, in the collection of Mr. J. H. Wilson, of Whitby, who kindly sent it for my inspection.

Nyctala acadica (*Gmel.*). Acadian Owl.

Doubtful, a native of North America.

Beverley, one (Milner, Zool., 1860, p. 7104).

158. Scops giu (*Scop.*). Scops Owl.

Accidental visitant from Southern Europe and Northern Africa, of extremely rare occurrence. First recorded as British from Yorkshire specimens.

Wetherby, one shot, spring of 1805, in the possession of Mr. Charles Fothergill, of York (Mont. Orn. Dict. Supp.).

Yorkshire, one in the possession of Mr. Foljambe, believed by him to have been shot in the county (Id.).

Womersley (Allis. *fide* Morris, 1844).

Ripley, near Harrogate, a pair of old and two young birds (Allis, *fide* Morris, 1844).

Eshton Hall, near Gargrave, one shot (Allis, 1844).

Boynton, near Bridlington, one shot, July, 1832 (Allis, *fide* Strickland).

Driffield, one shot about 1839 (Allis).

Sandhutton, one seen (Allis, 1844).

Egton Bridge, near Whitby, one shot, 1865 (Birds of Yorkshire, p. 52).

Scops asio (*L.*). Mottled Owl.

Accidental visitant from North America, of extremely rare occurrence.

Leeds, one shot in Hawksworth Wood, summer of 1852 (Hobson, Nat., 1855, p. 169 and plate). This I believe to be a genuine occurrence.

159. Bubo ignavus *Forst.* Eagle Owl.

Accidental visitant from Continental Europe, of extremely rare occurrence.

Yorkshire, one (Pennant, B. Zool., 1768, i. 157).

Horton, near Bradford, one, about 1824 (Denny, Leeds Cat.).

Harrogate, one taken alive in the summer of 1832 (Allis). Probably an escape.

Off Flamborough Head, one captured alive (Hawkridge, Wood's Nat., 1838, p. 155).

Greetland, near Halifax, one seen, Nov. 1845 (Birds of Yorkshire, p. 50).

Ilkley, one captured on Rombalds Moor, July, 1876 (Birds of Yorkshire, p. 51). Probably an escape.

Scarborough, one seen Oct. 30, 1879 (Clarke, Zool., 1880, p. 358).

The specimen recorded by Morris (B. Birds, 1851, i. 181) as shot at Clifton Castle, near Bedale, proves on investigation to be an escape from Hornby Castle.

160. **Athene noctua** (*Retz.*).   **Little Owl.**

Accidental visitant, from Continental Europe, of extremely rare occurrence. Has been recorded as a Yorkshire bird by Pennant (1768), by Berkenhout (1778), by Fothergill (1799), by Yarrell (1843), and by subsequent writers, and as late as Prof. Newton's edition of Yarrell (1871), but with an entire absence of particulars as to localities and dates. It is supposed to have occurred near Halifax (Leyland, 1828).

Sub-order *ACCIPITRES.*

Fam. **VULTURIDÆ.**

161. **Gyps fulvus** (*Gmel.*).   **Griffon Vulture.**

162. **Neophron percnopterus** (*L.*).   **Egyptian Vulture.**

Fam. **FALCONIDÆ.**

163. **Circus æruginosus** (*L.*).   **Marsh-Harrier.**

Casual visitant, of extremely rare occurrence; formerly nested in one or two localities. At the commencement of the present century this species bred commonly in the 'carrs' round Doncaster and Hatfield, and occasionally in the whin-beds near Bridlington. During the past thirty years, however, it has only occurred in four instances, viz. :—A female trapped at Cudworth, in April, 1869 ; a young male shot at Beverley, Oct. 13, 1871 ; an almost black specimen shot at Wassand, early in 1872; and one said to have been shot near Pocklington, in September, 1877.

164. **Circus cineraceus** (*Mont.*). **Montagu's Harrier.**

Casual visitant, of rare occurrence, chiefly as an autumn migrant. Formerly resident, and more widely distributed than either *C. æruginosus* or *C. cyaneus*. breeding in several localities, as on Thorne Waste, at Hackness near Scarborough, near Whitby, on Barden Moor in Wharfedale in 1860, and near Bridlington, as late as 1871.

165. **Circus cyaneus** (*L.*). **Hen-Harrier.**

Casual visitant, of very occasional occurrence, chiefly in autumn, but also in winter and spring. Formerly bred annually on the moorlands of North-eastern Yorkshire and the carrs near Doncaster, the last nest being found on the Danby Moors, about 1850.

166. **Buteo vulgaris** *Leach*. **Common Buzzard.**

Resident, but almost extinct, restricted to one—or at most two —pairs, nesting in the North-western Fells; observed rarely as an autumn or winter visitant. Formerly this bird was an abundant resident, especially amongst the crags of the North-western Fells, and also in the larger woods.

167. **Archibuteo lagopus** (*Gm.*). **Rough-legged Buzzard.**

Winter visitant, occurring annually, but in varying numbers, being scarce some years, whilst in others great flights arrive. Most frequent in the vicinity of the coast. These autumn immigrants are almost entirely immature birds, there being indeed only one instance on record of the occurrence of an adult. A pair bred for several years on the ground, amongst heather, near Hackness, Scarborough (More, Ibis, 1865).

168. **Aquila clanga** *Pall*. **Larger Spotted Eagle.**

169. **Aquila chrysaetus** (*L.*). **Golden Eagle.**

Casual visitant, of extremely rare occurrence.

Stockeld Park, near Wetherby, one shot Nov. 29, 1804 (Denny, Leeds Cat.).

Beningborough Park, near York, one trapped, Jan. 1838 (Wood's Nat., 1838, iii. p. 214).

Hunmanby, one shot July 24, 1844, now in the Scarborough Museum (Birds of Yorkshire, p. 2).

East Riding: 'Arthur Strickland reports that one has been killed' (Allis, 1844).

Kildale, one shot on Court Moor, Christmas, 1851, now in the collection of Capt. Turton, of Upsall Castle (Birds of Yorkshire, p. 2).

Skerne, near Driffield, a female in first year's plumage, shot Dec. 1861, now in the Norwich Museum (Cordeaux, Birds of Humber, p. 1).

Thornton, near Pickering, one shot in 1864 (Birds of Yorkshire, p. 3).

170. **Haliaetus albicilla** (*L.*). **Sea-Eagle.**

Casual visitant, of rare occurrence, most frequent in the winter and on the coast, but not confined to it. All the specimens known to have occurred were in immature plumage, but a bird shot at Castle Howard in 1841 had only two or three feathers of the tail tipped with black, having only these to lose in order to attain to mature plumage; it is now in the Leeds Museum.

171. **Astur palumbarius** (*L.*). **Goshawk.**

Casual visitant, of rare occurrence, in spring and autumn ; has been observed once or twice in winter, and is most frequent on the coast and its vicinity.

172. **Accipiter nisus** (*L.*). **Sparrow-Hawk.**

Resident, generally distributed, fairly numerous; observed on the coast as a regular autumn immigrant.

173. **Milvus ictinus** *Savigny.* **Kite.**

Casual visitant, of very rare occurrence ; formerly resident, and probably numerous, but there is positive evidence of its nesting in two instances only. These are Edlington Wood, where a pair were taken from the nest by Hugh Reid, about 1824 ; and Murton Wood, near Hawnby, where, early in the present century, Mr. Charles Harrison shot the female off the nest, also obtaining the male.

174. **Milvus migrans** (*Bodd.*). **Black Kite.**

175. **Nauclerus furcatus** (*L.*). **Swallow-tailed Kite.**
Accidental visitant from America, of extremely rare occurrence.

Hardraw Scarr, near Hawes, one captured alive, Sept. 6, 1805 (Newton's Yarrell, i. p. 104-5).

Other specimens are said to have been obtained near Helmsley, May 25, 1859; and in Bolton Woods, forty or fifty years ago (cf. Birds of Yorkshire, pp. 28, 29).

176. **Elanus cæruleus** (*Desf.*) **Black-winged Kite.**

177. **Pernis apivorus** (*L.*). **Honey-Buzzard.**
Casual visitant, of uncommon occurrence in spring and autumn, most frequent near the coast, and at the latter season.

178. **Falco candicans** *Gm.* **Greenland Falcon.**
Accidental visitant from Iceland, Greenland, Arctic North America, and Northern Asia, of extremely rare occurrence.

Sutton-on-Derwent, adult, Feb. 5, 1837 (Allis, Wood's Nat., 1837, p. 53).

Robin Hood's Bay, mature female, Nov. 25, 1854 (Roberts, Zool., 1855, p. 4588); now in the Scarborough Museum.

179. **Falco islandus** *Gmel.* **Iceland Falcon.**
Accidental visitant from Iceland and Southern Greenland, of extremely rare occurrence.

Guisborough, a young bird shot on the moors, March 1837 (Hogg, Zool., 1845, p. 1052).

Marston Moor, one in the collection of Mr. C. C. Oxley, said to have been obtained in December, 1826 or 1836 (Birds of Yorkshire, p. 10).

Upper Poppleton, near York, a young female shot Nov. 1860 (Graham, Zool., 1861, p. 7312). Now in the Leeds Museum.

Filey Brigg, a pair, one shot, Oct. 4, 1864 (Birds of Yorkshire, p. 10).

Whitby, a bird, probably of this species, found nailed up with 'other vermin' at Newton House in 1865 (Birds of Yorkshire, p. 11).

180. **Falco peregrinus** *Tunstall.* **Peregrine Falcon.**

Resident, now restricted to a pair—or perhaps two—breeding annually among the Fells of the North-West, and another on the cliffs of the coast, with an occasional pair in Cleveland; formerly it nested not uncommonly in suitable localities. In autumn and winter immature birds are not unfrequent on the coast, also occurring inland.

181. **Falco subbuteo** *L.* **Hobby.**

Casual visitant, of uncommon occurrence in summer, but has also been obtained several times in winter. As far as can be ascertained it has nested in the county on three occasions, at Rossington, near Doncaster (More, 1865), at Bishop Wood, near Selby, in 1869 (W. E. C.), and at Everingham Park, near Market Weighton (Boyes, 1875). Fifty years ago this species is mentioned as being far from uncommon in South Yorkshire.

182. **Falco æsalon** *Tunstall.* **Merlin.**

Resident, confined to the high western and north-eastern moorlands, over which it is thinly scattered during the breeding season. More generally distributed during the autumn and winter.

183. **Falco vespertinus** *L.* **Red-legged Falcon.**

Accidental visitant from Southern and Eastern Europe, of extremely rare occurrence.

Doncaster, a male shot in April, 1830; the first occurrence in Britain (Linn. Trans., xvii., p. 533).

Sheffield, one in the Museum, said to have been obtained in the district (Heppenstall, Zool., 1843, p. 247).

Easingwold, female (Allis, 1844).

Rossington, female (Allis, *fide* F. O. Morris, 1844).

Selby, female shot in Stainer Wood, May, 1844 (Zool., 1844, p. 654).

Humber mouth, female, Nov. 1864 (Boulton, Zool., 1865, p. 9415).

Bempton Cliffs, mature female, shot July 6th, 1865, now in the collection of Mr. J. Whitaker, of Rainworth (MS.).

Bempton, male, shot June 18, 1869 (Cordeaux, Birds of Humber, p. 5).

Egton Bridge, Whitby, 1876 or 1877 (Birds of Yorkshire, p. 17).

Scarborough, Mr. Roberts has preserved three specimens obtained there.

184. **Falco tinnunculus** *L.*   **Common Kestrel.**

Resident, generally distributed, the commonest of the Falconidæ. A regular autumn immigrant.

185. **Falco cenchris** *Naum.*   **Lesser Kestrel.**

Accidental visitant from Southern Europe, of extremely rare occurrence.

Wilstrop, near York, mature male, shot by Mr. John Harrison, of Wilstrop Hall, in the middle of November, 1867, now in the York Museum (Birds of Yorkshire, p. 21).

186. **Pandion haliaetus** (*L.*).   **Osprey.**

Casual visitant, of very rare occurrence; formerly regularly observed as a periodical visitant in spring and autumn, on its way to and from its northern breeding-haunts. More frequently observed in Eastern Yorkshire than elsewhere.

Order 3. STEGANOPODES.

Fam. PELECANIDÆ.

187. **Phalacrocorax carbo** (*L.*).   **Cormorant.**

Resident, breeding on the cliffs at Arncliffe near Saltburn, Runswick Bay, and those north and south of Robin Hood's Bay, also in limited numbers at Flamborough. At the latter station they formerly bred in some abundance, but were for some years banished, owing to constant persecution; a few have during the past two or three years returned to their former haunt—the result of the protection afforded by the Sea-birds Preservation Act. Not uncommon on the coast at other seasons, and occasionally observed inland.

188. **Phalacrocorax graculus** (*L.*).   **Shag.**

Periodical visitant to the coast, occurring in small numbers in spring and autumn, when on their way to and from their breeding stations; most numerous at the latter season,

when immature birds are not unfrequent at Flamborough.
Mr. A. Strickland informed Mr. Allis in 1844 that some
few years before it used to breed in considerable numbers
on the rocks off Flamborough Head.

189. Sula bassana (*L.*). Gannet.

Periodical visitant, common off the coast, and especially at
Flamborough, in the herring season.   Has occurred as a
straggler very far inland, but the examples so observed have
been in immature plumage.

Order 4. HERODII.

Fam. ARDEIDÆ.

190. Ardea cinerea *L.*   Common Heron.

Resident, local, but common.   Yorkshire heronries have
greatly decreased in number.   Those now in existence are
at Kildale-in-Cleveland, Newton Hall near Malton, Harewood Park near Leeds, Eshton Hall near Gargrave, and
Browsholme Hall near Clitheroe.   It nests singly and
irregularly in many parts of the county.

It is not uncommon on the coast in the autumn, when
immigrants arrive from the continent, and throughout the
winter.

191. Ardea purpurea *L.*   Purple Heron.

Accidental visitant from Southern Europe and Africa, of very
rare occurrence.

Flamborough, young bird shot in 1833 (Allis).

Lowthorpe, near Driffield, one, spring of 1847 (Morris, Zool.,
1849, p. 2591).

Temple Thorp, near Leeds, male, May 24, 1850 (Morris, B.
Birds, 1855, iv. 108).

Ruswarp, near Whitby, one, mature, shot, summer of 1850,
in the Whitby Museum (Stephenson, MS.).

Hornsea Mere, one, July, 1863, in the collection of Sir H.
S. Boynton (T. Boynton, MS.).

E

192. **Ardea alba** *L.* **Great White Egret.**

Accidental visitor from South-eastern Europe and Africa, of extremely rare occurrence.

Hornsea Mere, one, winter of 1821 (A. Strickland, Rep. Brit. Ass., 1838, p. 106).

Barnsley, one at New Hall, 1821, in the possession of Sir Joseph Radcliffe (Allis).

Scorborough, near Beverley, one about 1834 (Strickland, op. cit.); now in York Museum.

193. **Ardea garzetta** *L.* **Lesser Egret.**

Accidental visitant from Southern Europe and Africa, of extremely rare occurrence.

Hayburn Wyke, near Scarborough, one, Jan. 4, 1881 (Harper, Zool., 1881, p. 213).

194. **Ardea bubulcus** *Audouin.* **Buff-backed Heron.**

195. **Ardea ralloides** *Scop.* **Squacco Heron.**

Accidental visitant from Southern Europe and Africa, of extremely rare occurrence.

Askern, one, in the collection of Arthur Strickland (Allis, 1844).

196. **Ardetta minuta** (*L.*). **Little Bittern.**

Casual visitant, of uncommon occurrence, chiefly in summer. The following are the instances :—At Birdsall, near Malton, about 1842 ; at Thorp, near Bridlington, and at Doncaster prior to 1844; at Redcar, Sep. 20, 1852 ; at Hunslet, near Leeds ; at Harewood ; at Cottingham, near Beverley; at Cold Hiendley Reservoir, near Wakefield, Aug. 26, 1872 ; at Scarborough, Aug., 1873 ; Easington, near Spurn, May 25, 1874 ; at Ruswarp, near Whitby, May, 1877; a female at Scalby Beck, near Scarborough, Feb. 25, 1879 ; and at Filey Brigg in the winter of 1879.

197. **Nycticorax griseus** (*L.*). **Night-Heron.**

Accidental visitant from Southern and Eastern Europe and Africa, of very rare occurrence.

Cottingham, Hull, one, immature, 1837, in the collection of Sir H. S. Boynton (T. Boynton, MS.).

Birdsall, near Malton, one, May 21st, 1855 (D. Graham).

Whitby, one, autumn of 1861, in the collection of Mr. E. Corner (Stephenson, MS.).

Kirkby Misperton, near Malton, one, May, 1870, in' the collection of Mr. E. Tindall (Tindall, MS.).

198. Botaurus stellaris (*L.*). Bittern.

Casual visitor, in winter, of uncommon occurrence, being particularly numerous in severe seasons. In 1831 upwards of 60 specimens were obtained in the county. Doubtless it formerly nested in the carrs round Doncaster, and in Holderness.

199. Botaurus lentiginosus (*Mont.*). American Bittern.

Accidental visitor from North America, of extremely rare occurrence.

Slingsby, one shot at Kell's Springs, Dec. 4, 1871, in the collection of Sir John Crewe (Brigham, MS.).

Fam. CICONIIDÆ.

200. Ciconia alba *Bechst.* White Stork.

Casual visitor, of rare occurrence.

Howden, one, winter, 1798 (Fothergill, Orn. Brit., p. 7).

Bawtry, one about 1825 (Allis).

Bretton Park, Barnsley, one, March, 1831 (Allis).

Skipsea, one, in the collection of Arthur Strickland (Allis, 1844).

Wansford, near Driffield, one seen, spring of 1846 (Morris, Zool., 1846, p. 1501).

Riccall, male, May 18, 1848 (Milner, Zool., 1848, p. 2891).

Bessingby, near Bridlington, one, Sept. 18, 1856 (Roberts, MS.).

Withernsea, one, mature, Sept., 1869, in the collection of Mr. P. W. Lawton, of Easington (Cordeaux, Birds of Humber, p. 106).

201. Ciconia nigra (*L.*).   Black Stork.

Accidental visitant from Continental Europe, of extremely rare occurrence.

Market Weighton Common, one, Oct. 29, 1852 (B. R. Morris, Nat. iii. 19); now in the York Museum.

Fam. PLATALEIDÆ.

202. Platalea leucorodia *L.*   Spoonbill.

Casual visitant, of rare occurrence.

Staincross, one, July, 1833 (Allis).

Masham, one (Allis, 1844).

Tees Marshes, one killed some years ago (Hogg, Zool., 1845, p. 1172).

Horbury, near Wakefield, one, 1850 (Talbot, Birds of Wakefield).

Wilberfoss, near Pocklington, one, Aug. 2, 1851 (Milner, Zool., 1851, p. 3278).

Hornby Decoy, near Catterick, one, 1865; now in York Museum.

Richmond, one (Clark-Kennedy, Zool., 1868, p. 1135).

South Ferriby, one, spring, 1873 (Cordeaux, MS.).

Masham, one, 1877, in collection of Mr. C. C. Oxley (Oxley, MS.).

Fam. IBIDÆ.

203. Plegadis falcinellus (*L.*).   Glossy Ibis.

Accidental visitant from Central and Southern Europe and Africa, of extremely rare occurrence.

Easington, near Spurn, one, autumn of 1850, in the collection of the late Mr. Cuthbert Watson (Lawton, MS.).

Selby, a mature bird, at Brayton Bridge, last week of May 1874, in the collection of Mr. J. Jackson (Cordeaux, MS.; Matthewman, MS.).

Order 5. **ANSERES.**

Fam. **ANATIDÆ.**

204. Chenalopex ægyptiaca (*Gm.*). **Egyptian Goose.**

Has occasionally been observed, but is so frequently kept in a state of semi-domestication that it is now impossible to distinguish escapes from visitants.

205. Anser cinereus *Meyer.* **Grey-lag Goose.**

Casual visitant, of rare occurrence, in very small parties in the winter; has long ceased to breed in the Yorkshire carrs, where it was formerly abundant and resident.

206. Anser segetum (*Gm.*). **Bean-Goose.**

Winter visitant to the coast, but rare. Appears in late September or early October. Formerly immense flocks visited the Wolds during the day-time, returning to the coast at dusk.

207. Anser brachyrhynchus *Baill.* **Pink-footed Goose.**

Winter visitant, most abundant in eastern Yorkshire. The common wild goose of the county.

208. Anser albifrons (*Scop.*). **White-fronted Goose.**

Casual visitant in winter, and, although decidedly uncommon, is most frequent in severe seasons, on the coast and its vicinity, and has been occasionally observed far inland.

209. Bernicla brenta (*Pall.*). **Brent Goose.**

Winter visitant, common off the coast in most seasons, but especially so in severe ones, remaining sometimes until the beginning of April. Very occasional far inland, being a strictly marine species.

210. Bernicla leucopsis (*Bechst.*). **Bernacle Goose.**

Casual visitant, of rare and irregular occurrence on the coast. Three were seen at Spurn at Christmas, 1876. Has been observed far inland. Often confounded with *B. brenta.*

211. **Bernicla canadensis** (*L.*).   **Canada Goose.**

Has occasionally been observed and shot, but is now so
common in a semi-domesticated state on ornamental
waters, that it is impossible to discriminate between
escapes and visitants.

212. **Bernicla ruficollis** (*Pall.*).   **Red-breasted Goose.**

Accidental visitant from Northern Asia, of extremely rare
occurrence.

Wycliffe, one, winter of 1766 (Bewick, B. Birds, ii. 280).

Tees, two "seen of late years by the Tees." One of these
was afterwards shot on the Durham side, in Cowpen
Marsh (Hogg, Zool., 1845, p. 1178).

213. **Chen albatus** (*Cassin*).   **Cassin's Snow-Goose.**

Cygnus olor (*Gm.*).   **Mute Swan.**
Domesticated.

214. **Cygnus immutabilis** *Yarr.*   **Polish Swan.**

Accidental visitant (habitat unknown) of extremely rare
occurrence.

Off Bridlington Pier, a flock seen in 1844 by Mr. Arthur
Strickland, one of which was obtained (Allis).

Wilstrop, one shot by Mr. John Harrison. out of a party of
two or three, late autumn, about 1860 (Harrison, MS.).

215. **Cygnus musicus** *Bechst.*   **Whooper Swan.**

Winter visitant, occurring almost annually on the coast, in
numbers varying with the season, much less frequent
inland.   In severe winters large flocks are observed.

216. **Cygnus bewicki** *Yarr.*   **Bewick's Swan.**

Winter visitant, less frequent and numerous than *C. musicus*,
but still not uncommon in severe winters, when it is chiefly
observed in the estuary of the Humber.   Much rarer
inland.   Mr. Cordeaux remarks on the scarcity of immature birds (Birds of Humber, p. 158).

217. Tadorna cornuta *(Gm.).* **Common Sheldrake.**

Resident in extremely limited numbers, its breeding stations being confined to the sandhills bordering the estuaries of the Humber and Tees. In autumn and winter it is more common, being sometimes observed in large flocks, and stragglers are occasionally seen on inland waters.

218. Tadorna casarca *(L.).* **Ruddy Sheldrake.**

Accidental visitant from South and East Europe and North Africa, of extremely rare occurrence.

Cottingham, one killed some years ago, seen by H. B. Hewetson (MS.).

219. Anas boscas *L.* **Mallard.**

Resident, local but fairly abundant. In winter large numbers arrive from the north, usually in November, the species then becoming very generally distributed, and much more numerous. Formerly six decoys existed in the county, but the only one now worked is that at Hornby near Catterick, on the estate of the Duke of Leeds. In 1800 the decoys of Watton and Scorborough were destroyed by the Beverley and Barmston Drainage Scheme ; the former had an area of about 1000 acres of water, and has been known to yield nearly 400 ducks in one day. The two other Holderness decoys—Home and Meaux—ceased to exist about the same time. One on Coatham Marsh at the Tees mouth was in existence as late as 1872.

220. Chaulelasmus streperus *(L.).* **Gadwall.**

Casual visitant of very rare occurrence in winter. The following are the occasions :—At Swillington, prior to 1840; Stockton-on-Tees, one, Feb. 18, 1843 ; at Doncaster, a pair in the spring of 1844; on the Humber, a pair, March, 1851; at Hornby, one in the decoy, season of 1856–7 ; at Selby, in 1858 ; at Skerne near Beverley, a male, Jan. 31, 1871 ; Hempholme in Holderness, in 1876 ; and near York, four females seen, one shot, Dec. 15, 1880.

221. Spatula clypeata *(L.).* **Shoveller.**

Winter visitant, but not numerous, most frequent in Holderness and in the vicinity of the Humber ; of rare occurrence inland. Hewitson (Eggs of B. Birds, 1856, vol. 2, p. 400) says :—'Mr. Henry Milner tells me that it breeds on Hornsea Mere,' and I am informed by Mr. F. Boyes that the keeper told him he once found a nest, and that he has himself observed the bird there in the breeding season on more than one occasion.

222 **Querquedula crecca** (*L.*). **Common Teal.**

Resident, very local, breeding regularly on Strensall and Riccall Commons, and at Malham Tarn, and occasionally in other localities in the county. As a winter visitant, common, arriving sometimes as early as the middle of August.

223. **Querquedula circia** (*L.*). **Garganey Teal.**

Casual visitant in spring and autumn, more especially at the former season ; is most frequent in eastern Yorkshire, being a rare straggler far inland. Two in the collection of Mr. J. H. Gurney, jun., were obtained at Bridlington on the 1st and 2nd of June, 1868. Mr. M. Bailey possesses three which were obtained in autumn at Flamborough.

224. **Dafila acuta** (*L.*). **Pintail.**

Winter visitant, not uncommon on the Humber during severe seasons, but is rare far inland. Mr. Cordeaux tells me that mature males are scarce.

225. **Mareca penelope** (*L.*). **Wigeon.**

Winter visitant, common on the Humber and many inland waters. It has been noted that inland this bird is most abundant in mild seasons. Old females are comparatively rare.

Mr. Stephenson, of Whitby, informs me that a pair or two breed annually on the Fen Bog near that place, and that a female has been shot off the nest.

226. **Fuligula ferina** (*L.*). **Pochard.**

Resident, extremely local ; breeds annually in some numbers on Hornsea Mere, and formerly on the Mere at Scarborough ; has bred at Cold Hiendley Reservoir in 1861, and also in Craven. As a winter visitant is not uncommon.

227. **Fuligula rufina** (*Pall.*). **Red-crested Pochard.**

228. **Fuligula marila** (*L.*). **Scaup.**

Winter visitant, abundant on the Humber and coast, and not uncommon on many large inland waters. Arrives about the first week in November, leaving late in spring.

229. **Fuligula cristata** (*Leach*). **Tufted Duck.**

Winter visitant, common on the Humber and coast; less so on inland waters. One instance of its breeding at Malham Tarn in 1849 has been recorded (Cooke, Zool., 1849, p. 2879). It has also been observed to remain occasionally through the summer at Hornsea Mere, Sir W. Milner (Zool., 1854, p. 4441) surmising that it possibly bred there.

230. **Nyroca ferruginea** (*Gm.*). **White-eyed Duck.**

Casual visitant, of very rare occurrence.

Coatham Marsh Decoy, one taken Jan. 17, 1850 (Rudd, Zool., 1850, p. 2773).
Dalton, near Huddersfield, Dec., 1858 (Hobkirk's Nat. Hist. of Hudd., 1859, p. 145).
Coatham Marsh, pair seen, female shot, Oct. 3, 1878 (Rev. H. Smith, MS.).

231. **Clangula albeola** (*L.*). **Buffel-headed Duck.**

Accidental visitant from Northern America, of extremely rare occurrence.

Bessingby Beck, Bridlington, adult male, winter, 1864-5, in the collection of Mr. J. Whitaker (Whitaker, MS.; Cordeaux, Birds of Humber, p. 176).

232 **Clangula glaucion** (*L.*). **Goldeneye.**

Winter visitant, immature birds being not uncommon on the coast and in the Humber in severe seasons; old males are always rare. Inland it is frequently observed on extensive waters, and, as on the coast, is most abundant in severe winters.

233. **Cosmonetta histrionica** (*L.*). **Harlequin Duck.**

Accidental visitant from Northern Europe, Asia, and America, of extremely rare occurrence. Out of twenty-two British occurrences Mr. Dresser is of opinion that two only are referable to this species.

River Don, above Doncaster, a female shot; in the collection of H. E. Strickland (Allis, 1844).

Filey, young male, autumn, 1862; in the collection of Mr. Whitaker, of Rainworth (Roberts, MS.).

Hornby Decoy, male captured about 1860; in the collection of the late Mr. Geo. Savage, Keeper, Hornby Castle, Bedale. Has been seen by Mr. James Carter (Carter, MS).

234. Harelda glacialis (*L.*). Long-tailed Duck.

Winter visitant, immature birds being not uncommon off the coast, particularly at Flamborough Head and Bridlington Bay. The old birds only approach the shore in extremely severe weather. Rarely straggles inland, but has occurred as far west as York.

235. Somateria mollissima (*L.*). Eider Duck.

Casual visitant, observed off the coast in autumn and winter in limited numbers, chiefly immature birds. Its occurrence in the Humber is quite exceptional, and inland it has never been seen.

236. Somateria spectabilis (*L.*). King Eider.

Accidental visitant from Northern Europe, Asia, and America, of extremely rare occurrence.

One shot at Bridlington Quay early in August, 1850; was first recorded for Bedlington, in Northumberland, by Mr. Duff (Zool., 1851, p. 3036), and corrected by Mr. Hancock (B. of Northumberland and Durham, p. 159) on the authority of a letter received by him from Mr. Duff.

237. Somateria stelleri (*Pall.*). Steller's Duck.

Accidental visitant from Northern Europe and Asia, of extremely rare occurrence.

Filey (misprinted 'Filby'), male, assuming winter plumage, shot Aug. 15, 1845, by Mr. Curzon, and submitted to Mr. Yarrell for inspection (Bell, Zool., 1846, p. 1249); now in the collection of Mr. Thos. Boynton, of Ulrome Grange.

238. Œdemia fusca (*L.*). Velvet Scoter.

Winter visitant off the coast, rarely approaching the shore. Two instances are recorded of its occurring far inland: a mature male was shot at Clapham in Feb., 1841, and an example at Bentley near Doncaster prior to 1844. The 'Velvet Ducks' often reported inland are generally referable to *Œ. nigra*.

239. Œdemia nigra (*L.*). Common Scoter.

Winter visitant, common on the coast and in the Humber,
occasionally visiting inland waters. Arrives in immense
flocks in the autumn, and a few are found off the coast all
the year round. Mr. Cordeaux is of opinion that those
remaining during the summer are young of the preceding
year, not going north to breed. Occasionally large migratory flocks are observed far inland in spring.

240. Œdemia perspicillata (*L.*). Surf-Scoter.

241. Mergus merganser *L.* Goosander.

Winter visitant to the coast and its vicinity, in varying numbers, but not uncommon, especially in severe seasons, when
it is occasionally observed far inland on rivers and lakes.
Immature birds are most frequent, adults being considered
scarce.

242. Mergus serrator *L.* Red-breasted Merganser.

Casual visitant in winter, of rare occurrence, both on the
coast and inland; the recorded instances are, however,
somewhat too numerous to mention.

243. Mergus cucullatus *L.* Hooded Merganser.

Accidental visitant from North America, of extremely rare
occurrence.

Leeds, a pair obtained, in the collection of Mr. W. Christy
Horsfall (Gould, Birds of G. Britain, Part 10, 1866). These
specimens I have been unable to trace.

244. Mergus albellus *L.* Smew.

Winter visitant, not uncommon on the coast in severe seasons,
scarce inland. Old males are very rare, females and young
of the year predominating. Less frequent than formerly.

Series SCHIZOGNATHÆ.

Order 1. COLUMBÆ.

Fam. COLUMBIDÆ.

2.45. Columba palumbus *L.* Ring-Dove.

Resident, generally distributed, abundant. Flocks of immigrants appear in October and November, their numbers being dependent on the severity, or otherwise, of the season.

246. Columba livia *Bonnat.* Rock-Dove.

Resident, very local, but numerous where it occurs. Breeds in considerable numbers in the Flamborough range of cliffs. They are also reported to breed in several inland localities, as near Sheffield, where they are entirely absent in winter, near Masham, near Pateley Bridge, and near Richmond.

247. Columba œnas *L.* Stock-Dove.

Resident, local, but common in most localities where it occurs. This species is steadily increasing in numbers, and gradually diffusing itself over the county, being now frequent in localities where even ten years ago it was unknown. Mr. Allis (1844) stated that the only Yorkshire specimen he had seen was one in the York Museum, and that the only locality in which it was known to occur at that date was Sheffield, where they were reported as not unfrequent. Now the stock-dove breeds in all parts of the county, in localities too numerous to mention, but is least frequent in the north-west. At Flamborough it breeds plentifully in the sea-cliffs.

248. Turtur communis *Selby.* Turtle Dove.

Summer visitant, extremely local, and in very limited numbers. Breeds annually in Holderness and near Sheffield. A few are observed in Eastern Yorkshire during the autumnal migration. It is less frequent in the spring, but on April 15, 1878, fifteen were observed at Flamborough. In Western Yorkshire it is quite of exceptional occurrence.

249. **Ectopistes migratorius** (*L.*). **Passenger Pigeon.**

Accidental visitant from North America, of extremely rare occurrence.

Mulgrave, near Whitby, female shot, Oct. 12, 1876 (Hancock, Nat. Hist. Trans. North. & Durh., vol. v. p. 337 ; Zool., 1877, p. 180). Doubtless an escape.

Fam. PTEROCLIDÆ.

250. **Syrrhaptes paradoxus** (*Pall.*). **Pallas's Sand-Grouse.**

Accidental visitant from the Asiatic Steppes, of extremely rare occurrence.

A careful computation of the numbers which visited this county during the memorable irruption of the summer of 1863, shows that no less than 80 birds were observed in different localities, and that 24 of them were obtained. It is probable that, from the roving disposition of this species, identical birds would be recorded for more than one locality.

Order 2. GALLINÆ.

Fam. PHASIANIDÆ.

251. **Phasianus colchicus** *L.* **Pheasant.**

Semi-domesticated, resident, generally distributed, abundant.

252. **Caccabis rufa** (*L.*). **Red-legged Partridge.**

Resident in various parts of the county, but in extremely limited numbers, and only very occasionally shot. Mr. A. G. More (Ibis, 1865) mentions this bird as breeding ' very rarely ' in West Yorkshire. There is no evidence that I am aware of to show that this species has been introduced into the county.

253. **Caccabis petrosa** (*Gm.*). **Barbary Partridge.**

Accidental visitant from Northern Africa, of extremely rare occurrence.

Beverley, one about 1869, seen in the flesh by Mr. W. W. Boulton (Cordeaux, Birds of Humber, p. 81).

254. Perdix cinerea *Lath.* Partridge.

Resident, generally distributed, abundant. Mentioned as scarce at Halifax and in Upper Ribblesdale.

255. Coturnix communis *Bonnat.* Common Quail.

Summer visitant, breeding regularly in limited numbers, in Holderness and at Boston Spa, and irregularly in many other localities in the county. Has occurred occasionally in midwinter. Many records show that this species was formerly much more frequent, and Mr. A. Strickland informed Mr. Allis (1844) that they used to be taken in nets near Bridlington.

256. Ortyx virginianus (*L.*). Virginian Colin.

Accidental visitant from Eastern North America, of extremely rare occurrence.

Cottingham, male, 'a few years since,' in the collection of Mr. Boulton (Cordeaux, Birds of Humber, 1872, p. 83).

Fam. TETRAONIDÆ.

257. Lagopus mutus *Leach.* Common Ptarmigan.

258. Lagopus scoticus (*Lath.*). Red Grouse.

Resident, abundant on all the high moors, and in severe winters sometimes occurs as a straggler in the most unlikely localities. The Rev. H. H. Slater informs me—on the authority of his uncle, Mr. T. Horrocks, of Eden Brows, Carlisle—that towards the end of October every year there is a migration of packs of grouse from the Duke of Cleveland's moors, near High Force, in Upper Teesdale, to Mr. Horrocks' moors, at Alston, in Cumberland (a distance of twenty miles), where they remain until the end of the season, and then return to their own county. A large proportion of these migrants are hens, and they are different in size and plumage and readily discriminated from the Alston birds, being only two-thirds their size and weight, and their plumage more speckled and yellow.

259. **Tetrao tetrix** *L.* **Black Grouse.**

Resident, local, occurring chiefly near Sheffield, sparingly near Huddersfield, occasionally on the moors of Wensley-dale and Nidderdale, and near Richmond. At Lartington, in upper Teesdale, they have been introduced, and are now fairly numerous.

260. **Tetrao urogallus** *L.* **Capercaillie.**

Fam. **TURNICIDÆ.**

261. **Turnix sylvatica** (*Desf.*). **Andalusian Hemipode.**

Accidental visitant from Southern Europe and Northern Africa, of extremely rare occurrence.

Huddersfield, one near Fartown, April 7, 1865 (Gould, P.Z.S., 1866, p. 210).

Order 3. **GRALLÆ.**

Fam. **RALLIDÆ.** .

262. **Rallus aquaticus** *L.* **Water-Rail.**

Resident, local, far from numerous. Doubtlessly breeds annually in suitable localities in the county, but from the secluded nature of its habitats, and the skulking habits of the bird, its nest is very seldom found, young birds being more frequently met with. Winter visitant, immigrants arriving in September and October; the species is then more common, particularly so in severe seasons.

263. **Porzana maruetta** (*Leach*). **Spotted Crake.**

Resident, extremely local in its distribution and limited in its numbers; as a winter visitant very local but not un-common. A few pairs breed annually on the sedgy banks of the river Hull, at Beverley; and it has been known to breed occasionally near York and Doncaster.

264. Porzana bailloni (*Vieill.*). Baillon's Crake.

Casual visitant, of extremely rare occurrence.

Wensley, one on the banks of the Ure, April, 1807 (Fothergill in Whitaker's Richmondshire, 1823, i. 416); recorded as *Rallus pusillus* Pall., it is probably referable to this species. Mr. Dresser (Birds of Europe) cites this occurrence under *P. parva*, and gives the date as May 6th, 1807.

Huddersfield, one at Kirkheaton, May 29, 1874 (Palmer, Zool., 1874, p. 4159).

Goole, one killed a few years since is in the possession of Mr. Gunnec, of Thorne (Bunker, MS.).

265. Porzana carolina (*L.*). Carolina Crake.

266. Porzana parva (*Scop.*). Little Crake.

Casual visitant, of extremely rare occurrence.

Scarborough, one killed (W. C. Williamson, P.Z.S., 1836, p. 77).

Cantley, near Doncaster, one taken alive (Allis, 1844).

Aldwarke Bridge, above York, one flew into a coal-boat; now in the possession of Mr. Johnson, of Masham (Jno. Harrison, MS.).

267. Crex pratensis *Bechst.* Land-Rail.

Summer visitant, generally distributed and common, except in the manufacturing districts. Some seasons have been remarkable for its scarcity. Arrives early in May, departing in September.

268. Gallinula chloropus (*L.*). Moorhen.

Resident, generally distributed, common.

269. Fulica atra *L.* Common Coot.

Resident, generally distributed, and common, except in the manufacturing districts and the Western Fells, where it is local and not numerous. Breeds at Malham Tarn, 1250 feet above sea level.

Fam. **GRUIDÆ.**

270. Grus communis *Bechst.* **Common Crane.**

Accidental visitant from Northern Europe, of extremely rare occurrence.

York, one shot in 1797 (Fothergill, Orn. Brit., 1799, p. 7).

271. Grus virgo *(L.).* **Demoiselle Crane.**

Order 4. **LIMICOLÆ.**

Fam. **OTIDÆ.**

272. Otis tarda *L.* **Great Bustard.**

Accidental visitant from Continental Europe, of extremely rare occurrence; formerly resident in great numbers on the Wolds of Eastern Yorkshire, when in their virgin state as undulating barren sheepwalks.

The precise date of extinction is uncertain, but there is reason to believe that the last bird was killed at Reighton near Hunmanby, about the year 1830.

It is much to be regretted that the whole of the records of the existence in Yorkshire of so fine and conspicuous a bird should date subsequently to its extinction, and it is somewhat remarkable that no allusion to its presence in the county should be made by Pennant or other contemporary writers; probably this may be explained by the very abundance of the species. Even the records that exist are derived from memory, or based upon hearsay statements.

The materials available for treating of the past history of Yorkshire Bustards consist of—Mr. Arthur Strickland's account given in Allis's report on Yorkshire Birds, in 1844; notes by Mr. Henry Woodall, of North Dalton, and Mr. E. H. Hebden, of Scarborough, contributed to Morris' British Birds, in 1854; articles in the Zoologist for 1870 (pp. 2063, 2102, 2103); a letter from Sir Charles Anderson, of Lea, to Mr. John Cordeaux, dated Dec. 14, 1874; and letters to myself from Mr. Thos. Boynton, of Ulrome, Sir

F

C. W. Strickland, of Hildenley, and Mr. J. W. Woodall, of Scarborough. From such of these materials as have been published, the numerous statements given in books have been compiled.

At the northern extremity of the Wolds, the chief and last haunt of the Great Bustard seems to have been about Flixton, Hunmanby, and Reighton. It was here—as she informed Mr. Boynton—that the late Miss Charlotte Rickaby, of Bridlington Quay, when a girl, counted fifteen Great Bustards in a field while riding with her father from Bridlington Quay to Flamborough, early in the present century; and Sir C. W. Strickland informs me that his grandfather, Sir William Strickland, used to say that he could remember a flock of them on the Wolds between Reighton and Bridlington, of about five-and-twenty, and that the last of them was eaten at Boynton. A farmer living at Reighton in 1830 told Sir Charles Anderson that when he was a boy flocks of eight and ten together were found all over the district. Mr. Hebden states that to the best of his recollection it would be about 1811 that he first saw the five large Bustards on Flixton Wold, that number continuing there at least two years, when two were shot; the remaining three still continued on the same wold for at least one year, when two disappeared, leaving the solitary bird, which, after a length of time, was severely wounded by Sir Wm. Strickland's keeper, and found some days afterwards in a turnip field near Hunmanby, by the huntsman of the Scarborough Harriers, and secured. Mr. A. S. Bell adds that this bird was brought to Scarborough, and cooked at a supper given by the hunt (Zool., 1870, p. 2063). This, however, would hardly be the last Bustard, unless indeed the solitary individual survived its former companions for no less than fifteen years. Mr. J. W. Woodall informs me that about 1825 one was run over and killed between Folkton and Hunmanby. Sir Charles Anderson has a stuffed specimen, shot in 1825 at Hunmanby, and in 1828, while shooting on Mr. Osbaldeston's property at that place he saw a fine cock. This would, no doubt, be the identical bird seen in Grindale Field by Mr. John Milner, of Middledale, Kilham, he thinks about the year 1828, for—as he informed Mr. Boynton—it was some time after he left school in 1825, and at the time he was riding with his father, who died in 1830. Mr. Boynton was also told by the late Mrs. Metcalfe, of Bridlington Quay, that she and

her husband (who was vicar of Reighton, and died in
1834) were invited to dine at Boynton Hall with Sir Wm.
Strickland, the principal dish being a Great Bustard, which
Sir William in his note of invitation described as probably
the 'last of his race.' Sir Charles Anderson believes the
existence of the Great Bustard in Yorkshire ceased in
1832 or 1833, when the last hen bird was trapped on Sir
W. Strickland's estate at Boynton, near Bridlington.

Mr. Arthur Strickland, in the account which he furnished
to Mr. Allis in 1844, said that it used to be a constant
resident on the extensive wolds in the East Riding, but that
from the extension of tillage and the numerous enclosures
which had taken place during the half century, and from the
introduction of artificial crops, particularly saintfoin and
clover—which from being early cut often led to their
destruction—they rapidly decreased, and had then been for
some years quite extinct. About thirty years before [i.e.,
1814], when he first knew the district, the flock frequenting the part of the Wolds near Bridlington was reduced to
five or six, and appeared to remain at that standing for
some time, and he not unfrequently met with it when
riding about. It, however, soon became reduced, and it
was about fifteen years before [i.e., 1829] that the last was
killed at Reighton, since which none had been seen in the
neighbourhood. He believed those frequenting the Wolds
south of Driffield remained in existence some years longer,
but were then—at the time of his writing—totally exterminated.

In this last and somewhat off-hand statement, which he
does not substantiate, I am of opinion that Mr. Strickland
was mistaken, for, judging from the evidence which I am
able to quote, the birds on the north Wolds certainly
existed a few years later than those in the south.

The last Bustards which frequented the southern portion
of the Wolds were in the vicinity of North and South
Dalton. There is an egg—the only Yorkshire one known
to exist—in the Scarborough Museum, the note attached
to which states that it was found by Mr. James Dowker, at
North Dalton, in the East Riding, in 1810. This was presented to the Museum in March, 1840, by Dr. John Bury
(Fielden, Zool., 1870, p. 2063). Mr. John Wolley, the
eminent oologist, who saw the egg in 1843 and in 1850,
noted in his egg-book that it had been 'boiled with the
notion of preserving it' and was of 'bad colour' (Fielden,

tom. cit. 2102). Mr. H. Woodall informed Mr. Morris
that in 1816 or 1817 Mr. James Dowker, of North Dalton,
killed two near that place with a right and left shot, and
saw a third, Mr. Woodall believed, at the same time ; a
nest that had been forsaken was also found with one egg
in it, which is the one now in the Scarborough Museum.
One of the birds shot was presented to George the Fourth,
then Prince Regent. Mr. A. S. Bell (Zool., 1870, p. 2103)
adds that the other was cooked by Mr. Dowker, and that
in the previous year—which he states as 1809—five
Bustards were seen on the same moor, but were very wild,
and none shot. These dates disagree, but it is more than
probable that that of the label on the egg is the correct
one. Sir Charles Anderson also states that the Bustard
bred at Haywold [evidently the Hawold of the Ordnance
Map, situate above North Dalton] about 1810. In 1865
Mr. W. W. Boulton saw at Scorborough, the seat of Mr.
James Hall, two specimens which had been captured in
the East Riding—one, a female, was evidently a bird of
the year ; it was taken alive in the neighbourhood of Scorborough, about forty years before [i.e., about 1825], and Mr.
Hall had it tethered on his lawn ; the other, an old male,
Mr. Hall had forgotten the history of, but thought it was
taken not far from Doncaster, and certainly in Yorkshire.
(Zool., 1865, p. 9446). After Mr. Hall's death, his collection was sold, the male Bustard passing into the possession
of Mr. Thos. Boynton, and the female into that of Mr.
John Stephenson, of Beverley.

A pair—male and female—are preserved in the Black-
more Museum, at Salisbury, which were killed near
Malton, in 1825 (Thos. Norwood, of Salisbury, ' Country,'
Jan. 11, 1877, viii., 39).

The fine pair in the Scarborough Museum were purchased
from Mr. Reid, of Doncaster, and presented by Dr.
Murray, many years ago.

Since the date of its final extinction as a resident, the
Great Bustard—now become an accidental visitant—has
twice occurred in Yorkshire. A female was shot on
Rufforth Moor, near York, on the 22nd of Feb., 1861
(Allis, Zool., 1861, p. 7507), and is now in the York
Museum; and another female, just dead but still warm,
was picked up in the sea, near Bridlington Quay, on the
11th Nov., 1864 (Boulton, Zool., 1865, p. 9442).

273. **Otis tetrax** *L.* **Little Bustard.**

Casual visitor in winter, of rare occurrence. The localities from which it has been reported are:—Flamborough, two in the winter of 1814-15 ; Boythorpe, one early in 1839 ; Bolton Wood, near Bradford, about 1839 ; near Beverley, and one on the Wolds prior to 1844 (Allis) ; Goodmanham, a female, Jan. 19, 1854 ; Leven, near Beverley, mature female, Jan. 31, 1862 ; Allerston Marishes, near Scarborough, a pair, Oct. 1866 ; near Scarborough, one, a few years ago ; and North Burton, in 1868.

274. **Otis macqueeni** *J. E. Gray.* **Macqueen's Bustard.**

Fam. **ŒDICNEMIDÆ.**

275. **Œdicnemus scolopax** *(Gm.).* **Stone Curlew.**

Summer visitant to Eastern Yorkshire, where it is not only local in distribution, but very limited and fast decreasing in numbers. On the Wolds, where before their enclosure the bird was most numerous, it is now confined to a few localities in the north. From the Southern Wolds it is now all but banished. though only a very few years ago it was fairly abundant. Has bred occasionally on Levisham Moor, near Pickering ; on the Hambleton Hills, and near Scarborough. In 1844, Mr. Allis stated that it bred near Rossington, and other places in the vicinity of Doncaster. In Western Yorkshire it has been very rarely observed, and only as a straggler.

Fam. **GLAREOLIDÆ.**

276. **Glareola pratincola** *L.* **Common Pratincole.**

Accidental visitant from Central and Southern Europe, Africa, and Western Asia, of extremely rare occurrence.

Staxton Wold, near Scarborough, one, May, 1844 (Milner, Zool., 1848, p. 2023).

Bridlington, one in the winter of 1849-50 (Duff, Zool., 1850, p. 2771 ; and Hancock, Birds of Northumberland and Durham, 1874, p. 96). Mr. Duff originally recorded this specimen for ' Bedlington, in Northumberland ;' but in a letter to Mr. Hancock, informed him that the statement

in the Zoologist was a mistake, as the specimen was for-
warded to him from Bridlington, in Yorkshire.

Whitby, one, Oct. 19, 1871 (Simpson, Zool., 1871, p. 2870);
now in the Museum.

Fam. CHARADRIIDÆ.

277. Cursorius gallicus (*Gm.*).   Cream - coloured Courser.

Accidental visitant from the sandy plains of Northern Africa
and Western Asia, of extremely rare occurrence.

Wetherby, one, April, 1816 (Atkinson, Comp. B. Orn., p.
165).

One killed in 1825 by Lord Harewood's keeper (Allis; Gould,
Birds of G. Britain).

Holme, near Market Weighton, one killed in 1828, by Hon.
Chas. Stourton's keeper (Allis, *fide* Strickland).

278. Charadrius pluvialis *L.* Golden Plover.

Resident, breeding commonly on the moors of Cleveland and
the North-Western Fells; in the South-Western Moorlands
it is much less common, a few pairs breeding annually near
Halifax, Penistone, and Sheffield.   As a winter visitant it
is more general in its distribution, and is then observed in
flocks.   Arrives late in October and in November, but a few
old birds are sometimes observed early in August in summer
plumage.   Occasionally remains as late as the second week
in May.

279. Squatarola helvetica (*L.*). Grey Plover.

Winter visitant, observed abundantly in the spring and autumn
—chiefly on the coast—on its passage to and from its
breeding haunts in far North-east Europe.   Appears in the
autumn as early as August—the young birds arriving before
the old—but the majority arrive later, passing further south,
a few only remaining through the winter.   In May, it occurs
in flocks on the Holderness Coast, in all stages of plumage,
and leaves towards the middle and end of the month.
Inland it is but of occasional occurrence.

280. **Ægialitis cantiana** (*Lath.*). **Kentish Plover.**
Casual visitor, of extremely rare occurrence.

Ulrome, male and female shot by Mr. Thos. Boynton, May 25 and 28, 1869 (Boynton, Zool., 1869, pp. 1843-4.)
Flamborough : Mr. M. Bailey knows ' of one being shot here about 1857 ' (MS.).

281. **Ægialitis curonica** (*Gm.*). **Lesser Ringed Plover.**
Accidental visitor from Continental Europe, Asia, and Africa, of extremely rare occurrence.

Whixley, male, July 30, 1850 (Garth, Zool., 1850, p. 2953).
Others reported to me I believe to be referable to *Æ. hiaticula.*

282. **Ægialitis hiaticula** (*L.*). **Ringed Plover.**
Resident on the coast, breeding more or less commonly on sandy beaches, and also in arable fields near to the shore. At Whitby is only observed as a winter visitant. Inland it is only occasionally seen, but there is reason to believe that it nests, for I have seen several on the shingly shore of one of our inland reservoirs late in May. Large migratory flocks, consisting of young birds, arrive on the coast about the first week of August. The variety *intermedius* is occasionally observed in May in small flocks.

283. **Eudromias morinellus** (*L.*). **Dotterel.**
Periodical visitant, occurring in spring and autumn while passing to and from its breeding-stations ; most frequent on the coast, where it arrives regularly in the first week of May, sojourning for about a fortnight before taking its departure north. At this date it is also occasionally observed on the higher lands of the county. Much less frequently noticed in the autumn. Now occurs in much smaller numbers than formerly.

284. **Vanellus vulgaris** *Bechst.* **Lapwing.**
Resident, generally distributed, abundant. Less general in winter, when it frequents in flocks the lowlands and the coast. There are large arrivals of immigrants in autumn.

285. **Strepsilas interpres** (*L.*). **Turnstone.**
Winter visitant, on the coast. Arrives in considerable numbers during September, a few in August, the great majority

departing further south later in the autumn. The few remaining to pass the winter are joined in May by large flocks from the south, and leave along with them before the end of the month. For the last two years I have noted that about a score of non-breeding birds remain at Spurn throughout the summer.

236. Hæmatopus ostralegus *L.* Oystercatcher.

Winter visitant to the coast, local ; common in some localities, scarce in others. Occasionally frequents inland reservoirs. Sometimes observed on the coast as early as July, remaining till spring.

Fam. SCOLOPACIDÆ.

287. Recurvirostra avocetta *L.* Avocet.

Accidental visitant from continental Europe, of extremely rare occurrence.

Skipwith Common, two killed about 1824; one of them in the York Museum (Allis).

Spurn Point: Mr. Arthur Strickland informed Mr. Allis, in 1844, that he had known several to occur near the Spurn Lighthouse in spring some years before, but that he had heard of no recent occurrence.

Tees mouth, one shot in the winter of 1827-8 near the Tees (Hogg, Zool., 1845, p. 1172), a locality from which Mr. J. H. Gurney, jun. (Zool., 1876, p. 4765), records it as having occurred twice or three times.

The last instance in which the Avocet is known to have nested in Britain, was at the mouth of the Trent, about the year 1840. Mr. Hugh Reid, of Doncaster, informed Mr. A. G. More, in a letter dated June 1st, 1861, that eggs were taken on a sand island at the mouth of the river Trent about twenty years before. There was at the time a spring tide, which nearly covered the island, and the eggs were floating on the water. The man who took them shot one of the parent birds at the same time, and brought the eggs to Mr. Reid. The island had patches of grass growing on it, and there was always mud and warp about it—a likely place for the bird to feed on. The county boundary being at this place drawn in the centre of the river Trent, Yorkshire will share with Lincolnshire the honour of possessing the last British breeding-station of the Avocet.

288. **Himantopus candidus** *Bonnat.* **Black-winged Stilt.**

Accidental visitant from Southern and South-eastern Europe and Africa, of extremely rare occurrence.

Aike Carrs, near Beverley, two, old and young, shot by the keeper; formerly in the collection of the late Mr. Hall, of Scorborough, now in that of Mr. John Stephenson, of Beverley, who informs me that a woman living at Aike well remembers their being shot and shown to her.

289. **Phalaropus hyperboreus** (*L.*). **Red-necked Phalarope.**

Casual visitant in autumn and winter, of very rare occurrence. A specimen shot in Yorkshire, in the possession of Mr. Johnson, of Brignall, one of Ray's correspondents, was described and figured by Edwards (Nat. His. Birds, 1743?) as a 'Cock Coot-footed Tringa.' One was shot in the autumn of 1812, on Swinton Lake; one at Redcar, Nov. 22nd, 1851; one, mature, at Scarborough, in Dec., 1853; one at York, in summer plumage, in May, 1854; another at Scarborough, in Nov., 1854; and an immature male at Bridlington, Oct. 14, 1872. In 1844, Mr. Arthur Strickland informed Mr. Allis that it was occasionally met with on the Bridlington coast.

290. **Phalaropus fulicarius** (*L.*). **Grey Phalarope.**

Casual visitant in autumn and winter, of uncommon occurrence. Judging from the records, it appears to have been observed as frequently inland as on the coast.

291. **Scolopax rusticola** *L.* **Woodcock.**

Resident, known to breed annually in limited numbers in many woods in the county. Much better known as a winter visitant, arriving on the coast in October and November, sometimes in immense numbers, at others only a few being observed. Returns to the coast, for departure, during the first week of March.

292. **Gallinago major** (*Gm.*). **Double Snipe.**

Casual visitant, in autumn and winter, of uncommon occurrence.

The melanic variety, **sabini,** has been recorded as occurring twice in Wharfedale in August, 1820; one specimen at Otley on the 14th, and a second on the 17th at Denton Park (T. G[arnett], Loudon's Mag., 1835, p. 614).

293. **Gallinago cœlestis** (*Frenzel*). **Common Snipe.**

Resident, local, breeds in more or less numbers in all suitable localities. Immigrants arrive, often in immense numbers, late in October or early in November; and during the winter it is much more generally distributed, but very variable in its movements.

294. **Gallinago gallinula** (*L.*). **Jack Snipe.**

Winter visitant, common in suitable localities throughout the county. Usually arrives in October, departing in April; an exceptionally early occurrence has been recorded for the 20th of August.

295. **Limicola platyrhyncha** (*Temm.*).   **Broad-billed Sandpiper.**

Accidental visitant from Northern Europe, of extremely rare occurrence.

Hornsea Mere, one shot by T. Ellotson, April, 1863, now in the collection of Sir H. S. Boynton (T. Boynton, MS.).

296. **Tringa maculata** *Vieill.* **Pectoral Sandpiper.**

Accidental visitant from North America, of extremely rare occurrence.

Filey, one (Morris, B. Birds, 1854, iv. 316).
Teesmouth, one, Aug., 1853 (Rudd, Morris' Nat., 1853, p. 275).
Coatham, Redcar, one, Oct. 17, 1853 (Id.).

297. **Tringa fuscicollis** *Vieill.* **Bonaparte's Sandpiper.**

298. **Tringa alpina** *L.* **Dunlin.**

Resident, confined entirely to the high moorlands of the western border from north to south, over which it is scattered irregularly in most extremely limited numbers. Mr. Arthur Strickland informed Mr. Allis in 1844 that he had many years before taken both eggs and young on Stockton Common, near York. Occurs in vast numbers

as a winter visitant on the coast and the Humber,
especially the latter, where it arrives in August, and departs
in April and May. Occasionally occurs in limited numbers inland. A few remain on the coast during the
summer; these are probably young of the preceding year,
not nesting.

Mr. Cordeaux informs me that the small race of dunlin
(**Tringa schinzii** *Brehm*) occurs occasionally on the
Humber flats in May.

299. **Tringa minuta** *Leisl.*   **Little Stint.**

Casual visitant, of uncommon occurrence on the coast in
spring and autumn, most frequent at the latter season.
Specimens have been obtained in May in full summer
plumage. Has once or twice occurred inland.

300. **Tringa temmincki** *Leisl.*   **Temminck's Stint.**

Casual visitant, of extremely rare occurrence.

Scarborough, 'has been killed' (W. C. Williamson, P.Z.S.,
1836, p. 77).

Bridlington Quay, ' Mr. Boulton . . . once examined a
specimen shot . . . near Bridlington Quay' (Cordeaux, Birds of Humber, 1872, p. 137).

301. **Tringa minutilla** *Vieill.*   **American Stint.**

302. **Tringa subarquata** (*Güld.*).   **Pygmy Curlew.**

Periodical visitant in spring and autumn, uncommon on the
coast in August and September; still less common in spring.
A large and unusual flight appeared on the Humber on
August 30 and Sept. 1, 1873. A rare straggler inland.

303. **Tringa striata** *L.*   **Purple Sandpiper.**

Winter visitant, not uncommon on the coast, especially the
rocky portions. The young arrive early in September, old
birds in October. Has been observed late in April.

304. **Tringa canutus** *L.*   **Knot.**

Winter visitant, abundant on the coast, and especially on the
Humber, arriving usually in autumn ; the young in late
August or in September, followed by immense flocks of
old birds late in October or in November. Many of these

retire south on the approach of winter.   Immigrants on
their way north occur on the coast in April and May,
many in summer plumage.   Has occasionally appeared far
inland.

305. Machetes pugnax (*L.*).   Ruff.

Periodical visitant, in extremely limited numbers, observed
chiefly in Holderness and on the Humber, especially about
Paull, on their spring passage in May, and in autumn in
August and September.   A very rare straggler inland.

Until about the commencement of the present century
this species was abundant, and bred in the carrs of East
Yorkshire, on Skipwith Common near Selby, and also on
Hatfield Chase and the carrs about Doncaster, where Mr.
Hugh Reid—as he informed Mr. More —remembered
their breeding quite plentifully.   Mr. Arthur Strickland
informed Mr. Allis, in 1844, that before the drainage of the
carrs they used to be taken in considerable numbers in the
breeding season, but that he should doubt if any had bred
in the county within the half-century.

306. Calidris arenaria (*L.*).   Sanderling.

Winter visitant, common on the coast, arriving in July and
August, and leaving in May.   Of shore birds this is the last
to leave in spring and the first to arrive in autumn.   Far
inland it is rarely observed as a straggler.

307. Tryngites rufescens (*Vieill.*).   Buff-breasted Sand-piper.

308. Bartramia longicauda (*Bechst.*).   Bartram's Sand-piper.

309. Totanus hypoleucus (*L.*).   Common Sandpiper.

Summer visitant, breeding in more or less abundance on the
streams and reservoirs in the north-eastern, western, and
west-central portions of the county, but not in the lowlands.   Arrives in mid-April, leaving in August and
September.   In East Yorkshire it is observed during
migration, but it is probable that they may breed on the
streams about Driffield, Mr. F. Boyes informing me that
he has never failed to observe the birds there in the
summer months.

310. **Totanus macularius** (*L.*). **Spotted Sandpiper.**

An accidental visitant from North America, of extremely rare occurrence.

Tees, one shot, in the collection of Mr. John Grey (Hogg, Zool., 1845, p. 1173).

Bridlington, one said to have been seen, March 2nd, 1848 (Higgins, Zool., 1848, p. 2147).

Whitby, one, March 29, 1849 (Milner; Higgins; Zool., 1849, p. 2455–6).

311. **Totanus ochropus** (*L.*). **Green Sandpiper.**

Periodical visitant, in spring and autumn, when it is not uncommon in the Holderness drains; in a few instances has remained through both winter and summer, but there is no reliable evidence of its ever being bred. Most numerous in the autumn, arriving early in August.

312. **Totanus glareola** (*L*). **Wood-Sandpiper.**

Casual visitant, of rare occurrence.

Campsall, near Doncaster, one in the possession of H. Reid (Allis, 1844).

Beverley, three on the river, Aug. 4. 1878 (Boyes, MS.).

Kilnsea, Holderness, immature male shot, Sept. 7, 1878 (Clarke, Nat., 1879, p. 179). Another was seen by me the same day.

313. **Totanus flavipes** (*Gm.*). **Yellowshank.**

Accidental visitant from North and South America, of extremely rare occurrence.

Tadcaster, one, Oct. 1858 (Milner, Zool., 1858, p. 5958; Graham, Nat., 1858, p. 291).

314. **Totanus calidris** (*L.*). **Common Redshank.**

Resident, very local, breeds in limited numbers on Strensall Common, near Beverley, Pilmoor, Thorne Waste, and on Malham Tarn Moss (1250 feet); up to last year they also bred in some numbers on Riccall Common near Selby, which has since been drained and enclosed. In winter, a common shore bird on the Humber and Tees estuaries, large migratory flocks arriving in the early autumn.

315. Totanus fuscus (*L.*). Spotted Redshank.

Casual visitant, of extremely rare occurrence.

Braithwell Grange, near Doncaster, about 1828 (Allis).

Tees-mouth, one in the collection of W. Backhouse (Zool., 1846, p. 1261).

Kilnsea, near Spurn, immature male, shot Aug. 23rd, 1869 (Boyes, Field, Oct. 30, 1869).

Spurn, immature female shot Sept. 6th, 1876 (Cordeaux, MS.).

316. Totanus canescens (*Gm.*). Greenshank.

Periodical visitant, in spring and autumn, most frequent at the latter season, when immature birds are chiefly observed. A few are also occasionally noticed during the winter. Much rarer inland than on the coast and its vicinity.

317. Totanus solitarius (*Wils.*). Solitary Sandpiper.

318. Macrorhamphus griseus (*Gm.*). Red-breasted Snipe.

319. Limosa lapponica (*L.*). Bar-tailed Godwit.

Winter visitant, in limited numbers, on the coast. Common in autumn, returning from its breeding stations in September and October, gradually decreasing in numbers as winter advances, only a few remaining throughout that season. Arrives punctually on the coast about the 12th of May, when birds are to be seen in full summer plumage. Also occurs rarely far inland.

320. Limosa ægocephala (*L.*). Black-tailed Godwit.

Casual visitant of very rare occurrence, in autumn and winter, chiefly on the Humber muds. Formerly resident. 'The late Mr. H. Reid, of Doncaster, has frequently told me that the Black-tailed Godwit used, within his recollection, to breed on Hatfield Moor, in which locality he once found the young birds himself' (More, Ibis, 1865).

321. Numenius borealis (*Forst.*). Esquimaux Curlew.

322. Numenius phæopus (*L.*). Whimbrel.

Periodical visitant in spring and autumn, when it is common on most portions of the coast. Leaves for more northern breeding-haunts in May, returning in July and August. A pair or more have occasionally remained at Spurn through the summer. Sometimes observed inland. The account of its nesting in Yorkshire furnished to Mr. More by Mr. Thomas Gough is simply incredible.

323. **Numenius arquata** (*L.*). **Common Curlew.**

Resident, local, but breeding in more or less abundance on all the high moorlands, least numerous in the south. It retires from its breeding-haunts in August for the coast, where it remains during the winter, returning in April. A few are observed on the Humber muds all through the summer.

Order 5. GAVIÆ.

Fam. LARIDÆ.

Sub-fam. *STERNINÆ*.

324. **Sterna macrura** *Naum.* **Arctic Tern.**

Periodical visitant on the coast, in spring and autumn, on its way to and from its breeding-stations, and much the most numerous at the latter season. Of very rare occurrence inland.

325. **Sterna fluviatilis** *Naum.* **Common Tern.**

Periodical visitant on the coast in spring and autumn, passing to and from its breeding haunts, and, like *S. macrura*, most common in autumn, but occurs much more frequently inland.

326. **Sterna dougalli** *Mont.* **Roseate Tern.**

Casual visitant, of extremely rare occurrence. In 1844, Mr. Allis stated that Mr. H. Reid had shot it at Scarborough and Hornsea. 'Argus,' writing in the 'Field' (Jan. 13, 1877, p. 44), records five in the Tees bay from the 1st to 11th August, 1876.

327. **Sterna minuta** *L.* **Little Tern.**

Summer visitant, breeding in yearly-decreasing numbers at Spurn ; arriving in May, leaving in September. On the reservoirs near Wakefield, these birds appear every spring and autumn.

328. **Sterna bergi** *Lichtenstein.* **Rüppell's Tern.**

329. **Sterna caspia** *Pall.* **Caspian Tern.**

Accidental visitant from the shores of Continental Europe and Africa, of extremely rare occurrence.

Filey, one, early in September, 1874 (Willis, Field, Nov. 15, 1879, p. 684).

330. **Sterna anglica** *Mont.* **Gull-billed Tern.**

Accidental visitant from Southern Europe and Northern Africa, of extremely rare occurrence.

Leeds, a mature bird, which had been shot at and wounded on a mill reservoir, was brought alive to Mr. H. Denny in the last week of July, 1843 (Denny, Ann. & Mag. N.H., 1843, p. 297).

331. **Sterna cantiaca** *Gm.* **Sandwich Tern.**

Periodical visitant, not uncommon off the coast in autumn, on its passage south. One was shot at Filey on the 15th of December, 1875, an unusual date.

332. **Sterna fuliginosa** *Gm.* **Sooty Tern.**

Accidental visitant from Northern and Central America. of extremely rare occurrence.

Scarborough, one in the collection of Mr. Edward Tindall, shot at Scalby in 1863 (Tindall, MS.).

333. **Sterna anæstheta** *Scop.* **Smaller Sooty Tern.**

334. **Hydrochelidon hybrida** (*Pall.*). **Whiskered Tern.**

335. **Hydrochelidon leucoptera** (*Schinz*). **White-winged Black Tern.**

Accidental visitant from Southern Europe, of extremely rare occurrence.

Scarborough, one shot in 1860 is now in the collection of Mr. Edward Tindall (Tindall, MS).

Flamborough, a single mature bird was seen for some days in the spring of 1867, but not procured (Cordeaux, Birds of Humber, p. 197).

336. **Hydrochelidon nigra** (*L.*). **Black Tern.**

Periodical visitant to the coast and Humber, in spring and autumn, and not uncommon. During the present year (1881) a few have been noted at Spurn as late as the middle of June. Mr. Arthur Strickland informed Mr. Allis, in 1844, that it used to breed on some of the streams near Driffield, though it had ceased to do so some years before.

337. **Anous stolidus** (*L.*). **Noddy Tern.**

Sub-fam. *LARINÆ*.

338. **Xema sabinii** (*Sabine*). **Sabine's Gull.**

Accidental visitant from Arctic Asia and America, of rare occurrence.

Bridlington, female, Sept. 5, 1866 (Boulton, Zool., 1867, p. 543).

Bridlington, in full summer plumage, Aug. 10, 1872 (Gurney, jun., Zool., 1872, p. 3316).

Flamborough, one, Oct. 15, 1873, in the collection of Mr. J. H. Gurney, jun.

Bridlington, one, Oct. 14, 1875, in the collection of Mr. J. Whitaker (Whitaker, MS.).

Scarborough, one, immature, Nov. 7, 1878 (Roberts, Zool., 1878, p. 455).

Scarborough, one, immature, Nov. 1879, in the collection of Mr. Edward Tindall (Tindall, MS.).

339. **Rhodostethia rosea** (*Macgill.*). **Cuneate-tailed Gull.**

Accidental visitant from Arctic America, of extremely rare occurrence.

One killed near Tadcaster, Dec. 22nd, 1846 (Milner, Zool., 1847, p. 1694), but also described as shot at Milford-cum-Kirby [near Tadcaster] in Feb., 1847 (Charlesworth, Proc. Yorks. Phil. Soc., 1847, p. 33). Mr. Howard Saunders, who has seen the specimen, pronounced it to be in winter plumage (Field, Feb. 1875, p. 196).

340. **Pagophila eburnea** (*Phipps*). **Ivory Gull.**

Accidental visitant from high northern latitudes of Europe, Asia, and America, of extremely rare occurrence.

G

Scarborough, one shot 'many years ago' by Mr. C. Watson,
of York (Allis, 1844).

Filey, adult male, Aug. 1875 (Tuck, Zool., 1875, p. 4689).

Redcar, one shot, Nov., 1879 (Mussell, Field, Feb. 14, 1880).

Filey, adult male, autumn, 1880 (Backhouse, Friends' Nat.
Hist. Journal, 1881, p. 39).

341. Larus atricilla *L.* Laughing Gull.

Accidental visitant from the eastern coast of North America,
of extremely rare occurrence.

Filey, adult male, March, 1876 (Tuck, Zool., 1876, p. 4960).

342. Larus ridibundus *L.* Black-headed Gull.

Resident, extremely local in the breeding season. Common
on the coast in early spring, late summer, and autumn ;
less numerous in winter. Though formerly breeding in
several localities, its stations are now reduced to two—a
large colony on Thorne Waste, and a few pairs which have
this year (1881) revisited Strensall Common—a former
resort. Up to last year this bird bred plentifully on
Riccall Common—now enclosed, broken up, and drained.
Mr. F. S. Mitchell informs me that in 1860 a colony
appeared on the shores of a tarn on Newton Fell, and
deposited a large number of eggs, but these being all
taken, the gulls left the place and never returned.

343. Larus melanocephalus *Natt.* Adriatic Gull.

344. Larus ichthyaetus *Pall.* Great Black-headed Gull.

345. Larus minutus *Pall.* Little Gull.

Periodical visitant to the coast, in very limited numbers, in
autumn and winter, most frequent at the former season.
A female was obtained at Flamborough on the 13th of
July, 1868, in full summer plumage, but birds of the year
and old ones in winter dress are most frequent. In
February, 1870, after a terrific gale from the East, no less
than twenty-nine were obtained at Bridlington, nineteen
old and ten young birds.

346. Larus philadelphia *Ord.* Bonaparte's Gull.

347. **Larus canus** *L.* **Common Gull.**

Winter visitant, common on the coast throughout the autumn, winter, and early spring. A few immature birds frequent the Humber throughout the summer.

348. **Larus argentatus** *Gm.* **Herring Gull.**

Resident, common, having breeding-stations near Filey, Whitby, and Saltburn ; a few pairs also nest at Bempton.

349. **Larus fuscus** *L.* **Lesser Black-backed Gull.**

Winter visitant, common on the coast during the autumn, winter, and spring months ; a few immature birds remaining through the summer.

350. **Larus marinus** *L.* **Greater Black-backed Gull.**

Winter visitant, common, arriving on the coast in July, departing in spring for its more northern nesting-stations.

351. **Larus glaucus** *Fabr.* **Glaucous Gull.**

Winter visitant to the coast, immature birds being but of uncommon occurrence. Mature birds are rare, but in the severe winter of 1830 considerable numbers occurred, and several have appeared at Flamborough since 1872.

352. **Larus leucopterus** *Faber.* **Iceland Gull.**

Casual visitant to the coast in winter, when immature birds have been very rarely observed, but is possibly much overlooked. The adult bird has only once occurred in the county, as recorded by Yarrell in 1843 ; particulars are not now attainable.

Redcar, an immature bird, winter of 1854-5 (Rudd, Nat., 1855, p. 144).

Scarborough, immature bird, Jan. 15, 1867 (Knight, Zool., 1867, p. 637).

Flamborough, young bird, Oct. 12, 1867 (Cordeaux, Zool., 1867, p. 1010).

Bridlington, immature male, Dec. 15, 1870 (Boynton, Zool., 1871, p. 2488).

Whitby, one in the local collection at the Museum (Stephenson, MS.).

353. Rissa tridactyla (*L.*). Kittiwake.

Resident, breeding abundantly in the Flamborough range of
cliffs. Many retire south in autumn, though the species
is common on the coast in winter.

Sub-fam. *STERCORARIINÆ.*

354. Stercorarius catarrhactes (*L.*). Common·Skua.

Periodical visitant, occurring off the coast in autumn, and
most frequently off Flamborough Head ; comparatively
rare elsewhere. There is one record of its occurrence in
the spring, at Flamborough, on the 1st of March, 1868.
Has once been obtained inland, at York, 8th Oct., 1879.

355. Stercorarius pomatorhinus (*Temm.*). Pomatorhine Skua.

Periodical visitant, occurring off the coast in autumn. In
most years only one or two—in immature plumage—are
observed, principally at Flamborough. In October, 1879,
an irruption of considerable extent took place, both adults
and immature birds being extremely numerous on the
coast, and some were shot very far inland. Before the
date of this irruption, adult birds had been almost unknown
in the county.

356. Stercorarius crepidatus (*Banks*). Richardson's Skua.

Periodical visitant to the coast, in autumn, when it is numerous
off Flamborough, chiefly immature, old birds being uncommon. This species was extremely abundant with its
congeners in the great flight of 1879. Has occasionally
occurred as a straggler inland.

357. Stercorarius parasiticus (*L.*). Buffon's Skua.

Casual visitant in autumn, very much less frequent than *S.
pomatorhinus* and *S. crepidatus*, and although many—
mature and immature—were obtained during the memorable irruption of 1879, it was much less numerous than
they, and until then immature specimens only had been
obtained. Inland, one occurred at Strensall Common on
the 8th of October, 1879.

Order 6. TUBINARES.

Fam. PROCELLARIIDÆ.

358. Procellaria pelagica *L.* Storm-Petrel.

Winter visitant, not uncommon on the coast during severe weather, which drives them in from the open sea, and after gales of unusual severity they are often picked up dead or exhausted on the shores and in far inland localities.

359. Procellaria leucorrhoa *Vieill.* Leach's Petrel.

Casual visitant in winter, of rare occurrence. In the winter of 1831-2 many examples occurred far inland; four were picked up near York, one or two near Hull, one at Thirsk, and three or four near Halifax. Since then they have appeared several times, also far inland; one at Doncaster in 1837, one at Kirkhammerton in 1850-1, and one at Beverley in the autumn of 1854.

360. Oceanites oceanicus (*Kuhl*). Wilson's Petrel.

Accidental visitant from the Atlantic, of extremely rare occurrence.

Halifax, one shot at Southowram, late in Nov., 1874; now in the collection of Mr. Christopher Ward, to whom it was brought in the flesh (Ward, MS.).

361. Puffinus anglorum (*Temm.*). Manx Shearwater.

Periodical visitant, in spring and autumn, especially the latter season, when it is not uncommon off the coast, particularly at Flamborough.

362. Puffinus griseus (*Gm.*). Sooty Shearwater.

Casual visitant, of rare occurrence, in the winter. Mr. Dresser (Birds of Europe, parts 61 and 62, Aug., 1877) remarks on this species that it 'is difficult to discriminate the records of its occurrence in Great Britain, as it has so very generally been confused with *P. major*, from which it is clearly distinct.' Gould (Birds of Europe) and Yarrell both figured the present species as the Great Shearwater, and the former remarks (B. of G. Brit., vol. 5) 'that out of fifty

or eighty specimens which had come under his notice, not
more than three or four were' of 'the present species, all the
rest being Great Shearwaters, which shows that, as a rule, it is
much less numerous on our coasts than *P. major*—though on
the east coast the reverse would appear to be the case, at
least off the Yorkshire coast, for Mr. Cordeaux, who
believed the present species to be the young of *P. major*,'
says 'that most of the large Shearwaters which occur there
are referable to this species, and Mr. Boulton obtained
three near Flamborough in the autumn of 1866.'

363. Puffinus major *Faber.* Great Shearwater.

Casual visitant, of rare occurrence, in autumn and winter.
Mr. M. Bailey, of Flamborough, informs me that he shot a
fine adult on the 10th Jan., 1874.

After what Mr. Dresser has said, as quoted under *P. griseus*,
great difficulty arises in giving instances of the occurrence
of that and the present species, the records being inextri-
cably entangled, and it would be an impossibility, without
examining the specimens, to assign them to either form.

364. Fulmarus glacialis (*L.*). Fulmar.

Casual visitant to the coast, in autumn and winter, of rare
occurrence. Often seen in large numbers at sea off the
coast.

365. Œstrelata hæsitata (*Kuhl*). Capped Petrel.

366. Bulweria columbina (*Moq.-Tand.*). Bulwer's Petrel.

Accidental visitant from the Atlantic, of extremely rare
occurrence.

Tanfield, one picked up dead on the banks of the Ure, May
8, 1837; in the collection of Colonel Dalton (Yarrell, 1843,
iii. 514).

Scarborough, one, in the spring of 1849 (Higgins, *fide*
Graham, Zool., 1849, p. 2569). This record is eminently
unsatisfactory, from the absence of details necessary to
substantiate the occurrence of so rare a bird.

These appear to be the only recorded occurrences for the
British Isles and the European Continent, Madeira and
the Canaries being its only known localities.

Order 7. **ALCÆ.**

Fam. **ALCIDÆ.**

367. Alca torda *L.* Razorbill.

Resident, breeding in considerable numbers on the Flamborough cliffs. It arrives at the breeding stations early in the year, departing with its young late in July. After continued rough weather in winter, I have observed great numbers washed ashore dead on the coast of Holderness.

Alca impennis *L.* Garefowl.

368. Lomvia troile (*L.*). Common Guillemot.

Resident, breeding in vast numbers on the Flamborough cliffs, beginning to arrive there in May and June, leaving its breeding haunts in the middle of August. Remains off the coast through the winter.

369. Lomvia bruennichi (*Sabine*). Brünnich's Guillemot.

370. Uria grylle (*L.*). Black Guillemot.

Casual visitor, occasionally seen off the coast in spring, autumn, and winter, more frequently off Flamborough Head than elsewhere. Mr. Arthur Strickland, in 1844, informed Mr. Allis that thirty years before he had killed a specimen in summer plumage out of a small flock at Flamborough, in the height of the breeding season; whether they were breeding he could not say.

371. Mergulus alle (*L.*). Little Auk.

Winter visitor, not uncommon off the coast, occasionally driven in shore and far inland by stress of weather.

372. Fratercula arctica (*L.*). Puffin.

Resident, nesting in May and June, in immense numbers, on the Flamborough Cliffs, retiring far out to sea in the winter, at which season only a few are found off the coast. Of the Flamborough sea-fowl, this is the last to arrive at its breeding-station, which it leaves in mid-August.

Order 8. PYGOPODES.

Fam. COLYMBIDÆ.

373. Colymbus glacialis *L.* Great Northern Diver.

Winter visitant, not uncommon, but still not numerous. Most frequent on the coast, but also occurs inland, though rarely. Arrives in September, leaves early in spring.

374. Colymbus arcticus *L.* Black-throated Diver.

Winter visitant, of uncommon occurrence, on the coast and inland. Appears to be rather more frequently observed inland than *C. glacialis*, whilst it is much the rarer of the two on the coast.

375. Colymbus septentrionalis *L.* Red-throated Diver.

Winter visitant, common on the coast, and regularly appears at that season on many inland waters, but not numerously. May be found off the coast all the year round—probably young of the previous year not breeding.

Fam. PODICIPITIDÆ.

376. Podiceps cristatus (*L.*). Great Crested Grebe.

Resident, about half-a-dozen pairs breeding regularly at Hornsea Mere; the late Mr. Hugh Reid informed Mr. More in 1865 that it occasionally bred in the West Riding. Those nesting at Hornsea retire to the sea and Humber on the approach of severe weather. Also a winter visitant, observed both inland and on the coast, but far from commonly.

377. Podiceps griseigena (*Bodd.*). Red-necked Grebe.

Winter visitant, uncommon, but not unfrequent in severe seasons, when it occurs both on the coast and inland, from September to March. Examples have occurred in almost full summer plumage.

378. Podiceps auritus (*L.*). Sclavonian Grebe.

Winter visitant, not uncommon on the coast and inland, from October to sometimes as late as the end of April. With the exception of the Dabchick (*P. fluviatilis*) Mr. Cordeaux considers this to be the most numerous of the Grebes visiting the Humber. An adult female, in full summer plumage, was shot at Flamborough on the 29th of October, 1874.

379. Podiceps nigricollis (*C. L. Brehm*). Eared Grebe.

Casual visitant, of very rare occurrence.

Huggate, one, Dec. 8th, 1849 (Graham, Zool., 1850, p. 2747).

Scarborough, one, Feb., 1855 (Roberts, Zool., 1855, p. 4660).

Bubwith, adult male, 1854 (Boynton, MS.).

Wakefield, one at Thornes, Feb., 1861 (Talbot, Birds of Wakefield, p. 31).

Hull, a female, Feb. 20, 1864 (Boulton, Zool., 1864, p. 9048).

Ripon, one, in the possession of Mr. Parkin, was shot close to the city (Slater, MS.).

380. Podiceps fluviatilis (*Tunstall*). Little Grebe.

Resident, generally though thinly distributed. Decidedly less frequent on the higher lands of the county, though a pair' or so breed at as high an elevation as Malham Tarn—1250 feet above sea level. Also a winter visitant, arriving in October.

REPTILES

AND

AMPHIBIANS.

WM. DENISON ROEBUCK.

Class 3. REPTILIA.

Sub-class *SQUAMATA.*

Order **OPHIDIA.**

Family **COLUBRIDÆ.**

1. **Tropidonotus natrix** (*L.*). **Common Snake.**
Generally distributed in the lowland districts, but decidedly local, being numerous in some localities, scarce in others.

2. **Coronella lævis** *Lacép.* **Smooth Snake.**

Fam. **VIPERIDÆ.**

3. **Vipera berus** *L.* **Viper. Adder.**
Generally distributed, and common on moorlands and lowland heaths, also found in some woods and on waste lands.

Order **LACERTILIA.**

Fam. **LACERTIDÆ.**

4. **Lacerta vivipara** *Jacq.* **Common Lizard.**
Generally distributed, numerous on dry sandy heaths and moors, also on commons and waste places.

5. **Lacerta agilis** *L.* **Sand Lizard.**

6. **Lacerta viridis** *L.* **Green Lizard.**

Fam. **SCINCIDÆ.**

7. **Anguis fragilis** *L.* **Slow-worm. Blindworm.**
Generally distributed, not uncommon in old quarries, heaths, and dry sandy places generally.

Sub-class *CATAPHRACTA*.

Order CHELONIA.

Fam. CHELONIDÆ.

8. **Dermatochelys coriacea** (*L.*). **Leathery Turtle.**

Accidental visitant from the tropical and sub-tropical portions of the Atlantic and Mediterranean seas, of extremely rare occurrence.

Pennant, in his British Zoology, speaks of a 'Tortoise' that was taken off the coast of Scarborough in 1748 or 1749 ; this is quoted by Bell (British Reptiles, 1839, p. 15) in connection with the present species, though the evidence of specific identity appears to be entirely wanting.

A more satisfactory record is that of one in Bridlington Bay, where it was caught in the herring-nets on the evening of October 25, 1871. It measured eight feet in length, and the same between the tips of the flappers, and was estimated to weigh upwards of 1000 lbs. (Alwin S. Bell, Zool., 1872, p. 2907).

9. **Chelone imbricata** (*Schweigg.*). **Hawk's-bill Turtle.**

Accidental visitant from tropical seas, of extremely rare occurrence.

Off Redcar, a large specimen was found floating dead on the sea in the summer of 1849 (Rudd, Zool., 1850, p. 2707).

Class 4. AMPHIBIA.

Order BATRACHIA URODELA.

Fam. SALAMANDRIDÆ.

1. **Triton cristatus** *Laur.* **Great Crested Newt.**

Generally distributed, but more local and less numerous than the common Smooth Newt.

2. **Triton tæniatus** (*Schneid.*). **Smooth Newt.**

Generally distributed and very abundant.

3. **Triton palmipes** (*Latr.*). **Palmated Newt.**

Found near Clitheroe, Huddersfield, Barnsley, Leeds, Ilkley, Topcliffe, and Whitby. No doubt occurs also in other localities, but it is not usually distinguished from *T. tæniatus.*

Order BATRACHIA ANURA.

Fam. BUFONIDÆ.

4. **Bufo vulgaris** *Laur.* **Common Toad.**

Generally distributed and abundant.

5. **Bufo calamita** *Laur.* **Natterjack Toad.**

Only known in one locality—Mytton, on the Lancashire border. It has there been noticed by Mr. Thos. Altham for thirty years, but is not common, and is known by the name of 'harrow-toad.' This species I have the pleasure of adding to the Yorkshire fauna on the authority of Mr. F. S. Mitchell.

Fam. **RANIDÆ**.

6. Rana temporaria *L.* Common Frog.
Universally distributed and extremely abundant.

7. Rana esculenta *L.* Edible Frog.

FISHES.

WM. EAGLE CLARKE;

WM. DENISON ROEBUCK.

Class 5. PISCES.

Order 1. CHONDROPTERYGII.

Sub-order *PLAGIOSTOMATA.*

Division SELACHOIDEI.

Family **CARCHARIIDÆ.**

1. Carcharias glaucus (*L.*). Blue Shark.

Has frequently been caught off Whitby, and occasionally off other parts of the coast.

2 Galeus canis *Bonap.* Common Tope.

This species is reported to us as migrating from the shore to the open sea. It is common at Scarborough and in Bridlington Bay, but is not reported from other localities.

3. Zygæna malleus (*Risso*). Hammer-headed Shark.

4. Mustelus vulgaris *Müll. & Henle.* Smooth Hound.

Common in Bridlington Bay and off Scarborough. Not reported elsewhere.

Fam. LAMNIDÆ.

5. Lamna cornubica (*Gm.*). Porbeagle.

Occasional visitant, has been caught off Scarborough and Whitby. This species is described by some authors under the name of ' Beaumaris Shark.'

6. **Alopecias vulpes** (*Gm.*). **Fox Shark. Thrasher.**

Accidental visitant from the Atlantic and the Mediterranean Sea, of rare occurrence.

Scarborough, one seen, Sept. 1854 (Briggs, Zool., 1854, p. 4513).

Bridlington, one, Oct. 15, 1868, 12 feet long ; now in the Leeds Museum.

Whitby, one caught some years ago (Stephenson, MS.).

Redcar, one washed ashore near the Tees mouth in Oct. 1879, five feet in length (Nelson, MS.).

7. **Selache maxima** (*Gunner*). **Basking Shark.**

Supposed to have occurred near Scarborough on two occasions, but the evidence as to identification is insufficient.

Fam. NOTIDANIDÆ.

8. **Notidanus griseus** (*Gm.*). **Grey Notidanus.**

Fam. SCYLLIIDÆ.

9. **Scyllium canicula** (*L.*). **Small-spotted Dogfish.**

Resident, not uncommon along the coast. This species is the 'Nurse-Hound' of Couch. At Redcar it is known as 'Sea-Nurse.'

10. **Scyllium stellare** (*L.*). **Large-spotted Dogfish.**

The only authority for including this species is its enumeration in Dr. Murray's Scarborough list (1832).

11. **Pristiurus melanostomus** *Bonap.* **Black-mouthed Dogfish.**

Fam. SPINACIDÆ.

12. **Acanthias vulgaris** *Risso.* **Picked Dogfish.**

Resident, abundant. The common dogfish of the Yorkshire coast. At Redcar this is called 'Sea-Dog.'

13. **Læmargus borealis** (*Scoresby*). **Greenland Shark.**
Accidental visitant from Arctic seas and the North Atlantic, of rare occurrence.

Dogger Bank, two in Feb., 1866 (Cordeaux, Zool., 1866, p. 230).

Scarborough, two specimens brought in two years ago, and sent to the Oxford Museum (Woodall, MS.).

Whitby (nine miles off), one, March 4th, 1880; now in Whitby Museum (Stephenson, MS.).

Whitby (seven miles off), one captured Ap. 6, 1881 (Stephenson, MS.).

14. **Echinorhinus spinosus** (*Gm.*). **Spinous Shark.**
Accidental visitant from the Atlantic Ocean and Mediterranean Sea, of rare occurrence.

Filey Bay, one in the summer of 1830, figured by Yarrell (Yarrell, B. Fishes).

Bridlington Bay, one on the 11th August, 1838 (Arthur Strickland, Rep. Brit. Ass., 1838, p. 107; Yarrell, B. Fishes, third edition, vol. 2, p. 530).

Scarborough, one, June, 1853 (Murray, Morris's Nat., 1853, p. 277).

Fam. RHINIDÆ.

15. **Rhina squatina** (*L.*). **Angel-Fish. Monk.**
Occasionally captured off the Yorkshire coast. Is common on the Dogger Bank.

Div. BATOIDEI.

Fam. TORPEDINIDÆ.

16. **Torpedo hebetans** *Lowe.* **Cramp-Ray. Torpedo.**

Fam. RAJIDÆ.

17. **Raja clavata** *L.* **Thornback Ray.**
Resident, abundant.

18. **Raja maculata** *L.* **Homelyn Ray.**

19. **Raja radiata** *Donov.* **Starry Ray.**

The only evidence for the inclusion of this species in the Yorkshire fauna is its enumeration by Meynell in his list as one of the rarest of the Rajidæ occurring on the coast (Rep. Brit. Ass., 1844).

20. **Raja circularis** *Couch.* **Sandy Ray.**

Resident, only reported from three localities. Mentioned as abundant in Bridlington Bay, frequently met with at Scarborough, and once at Whitby.

21. **Raja batis** *L.* **Common Skate.**

Resident, abundant. Mr. Bailey considers that off Flamborough they have decreased in number. One mentioned by Mr. Cordeaux, measuring seven feet two inches in length and five feet eight inches across, weighed fifteen stones.

The variety **intermedia**, or **Flapper Skate,** has once or twice occurred at Whitby (Stephenson, MS.).

22. **Raja marginata** *Lacép.* **Bordered Ray.**

23. **Raja lintea** *Fries.* **Sharp-nosed Ray.**

Resident, taken not uncommonly off Whitby and in the North Sea.

24. **Raja fullonica** *L.* **Shagreen Ray.**

Resident, but only occasionally caught. Pennant's record of the occurrence of one at Scarborough has been copied by most subsequent writers, and also his statement that the fishermen there call it 'Whitehause.' Was formerly taken in the stake-nets near Bridlington; and has of late years been caught at Whitby, where it is known as the ' French Ray,' a name also in vogue at Grimsby.

25. **Raja vomer** *Fries.* **Long-nosed Skate.**

Resident, occasionally met with near Scarborough, and has been caught at Whitby. In the stake-nets near Bridlington, two were caught some years ago. Reported as common in the North Sea.

Fam. **TRYGONIDÆ.**

26. Trygon pastinaca (*L.*). Sting Ray.

Resident, occasionally caught off Whitby and in Bridlington Bay.

Fam. **MYLIOBATIDÆ.**

27. Myliobatis aquila (*L.*). Eagle Ray.

Visitant, of extremely rare occurrence. Was first recorded as British by Pennant, who states that one was caught off Scarborough, and the tail brought to Mr. Travis, of that place, by the fisherman who caught it.

28. Dicerobatis giornæ (*Lacép.*). Horned Ray. Ox Ray.

Sub-order *HOLOCEPHALA.*

Fam. **CHIMÆRIDÆ.**

29. Chimæra monstrosa *L.* Northern Chimæra.

Occasionally taken in the North Sea by the Grimsby smacks, and in all probability occurs within the latitude of the Yorkshire coast.

Order 2. **GANOIDEI.**

Sub-order *CHONDROSTEI.*

Fam. **ACIPENSERIDÆ.**

30. Acipenser sturio *L.* Sturgeon.

Not uncommon off the coast, also at the mouth of the Tees, in the Humber, and in the tidal reaches of its tributary rivers, which it ascends about June, its upward progress being, however, barred by the first dam or weir. Formerly —before the erection of these obstructions—it has been known to ascend the Ouse as far as Boroughbridge, and the Wharfe to Boston Spa and Arthington. Near Bridlington, it has been several times caught in the stake-nets.

Sub-class 2. *TELEOSTEI.*

Order 1. **ACANTHOPTERYGII.**

Div. *ACANTHOPTERYGII PERCIFORMES.*

Fam. **PERCIDÆ.**

31. **Perca fluviatilis** *L.* **Perch.**

Freshwater resident, common, generally distributed in the middle and lower waters of the rivers, but not found in them at a greater elevation than 600 feet ; entirely absent from the Cleveland streams. In ponds and canals throughout the county—introduced. Occurs in Malham Tarn, at an elevation of 1250 feet, where Dr. Whitaker (History of Craven) says they become blind and black. We are, however, informed by Mr. Walter Morrison that as regards their being black, such is not the case, though they are sometimes caught blind.

32. **Labrax lupus** (*Lacép.*). **Basse.**

Occasionally taken during the summer months off Whitby and Scarborough.

33. **Acerina cernua** (*L.*). **Ruffe. Pope.**

Freshwater resident, not uncommon ; irregularly distributed in the rivers of the lowland districts, and has also been introduced into canals and reservoirs. Formerly in the Calder and Lower Aire, and in some canals, but now exterminated by pollution. Generally known as 'Tommybarr.'

34. **Serranus cabrilla** (*L.*). **Smooth Serranus.**

35. **Serranus gigas** (*Brünnich*). **Dusky Serranus.**

36. **Polyprion cernium** *Val.* **Stone Basse.**

37. **Dentex vulgaris** *C. & V.* **Sparus. Dentex.**

Doubtful. Said to have occurred at Scarborough (Murray, 1832).

Fam. MULLIDÆ.

38. **Mullus barbatus** *L.* **Red Mullet.**
The variety **surmuletus** *L.*, or **Striped Surmullet**, is a casual visitant, occurring in summer at Redcar, Whitby, and in Bridlington Bay.

Fam. SPARIDÆ.

Group *CANTHARINA*.

39. **Cantharus lineatus** (*Mart.*). **Black Sea-Bream.**
Occasionally captured in Bridlington Bay, and also off Flamborough. Mr. Cordeaux saw one or two examples taken in the autumn about ten years ago, on long lines—baited with mussels—on rocky ground off Specton Cliffs: the fishermen called it 'Old Wife.'

40. **Box vulgaris** *C. & V.* **Bogue.**

Group *PAGRINA*.

41. **Pagrus vulgaris** *C. & V.* **Braize. Becker.**

42. **Pagrus auratus** (*L.*). **Gilthead.**

43. **Pagellus centrodontus** (*De la Roche*). **Common Sea-Bream.**
Resident, caught not uncommonly along the coast during the summer months. Known at Redcar as the 'Sea-hen.'

44. **Pagellus bogaraveo** (*Brünn.*). **Spanish Bream.**

45. **Pagellus owenii** *Günth.* **Axillary Bream.**

46. **Pagellus acarne** (*Cuv.*).

47. **Pagellus erythrinus** (*L.*).

Fam. SCORPÆNIDÆ.

48. **Sebastes norwegicus** (*Müll.*). **Bergylt. Norway Haddock.**
Has occurred frequently off the coast, at great depths. Was first mentioned by Pennant (B. Zool., 1770) as taken near Scarborough.

Div. *ACANTHOPTERYGII SCIÆNIFORMES.*

Fam. **SCIÆNIDÆ.**

49. **Sciæna aquila** (*Lacép.*). **Maigre.**
Accidental visitant from the Mediterranean Sea, of extremely rare occurrence.

> Redcar, a fine specimen, measuring five feet one inch in length, was found on the coast between Redcar and the Tees mouth, Dec. 24th, 1849 (Rudd, Zool., 1850, p. 2709).
>
> Flamborough (about seven miles off), one taken Aug. 25, 1873 (Boynton, MS.).

50. **Umbrina cirrhosa** (*L.*). **Umbrina.**

Div. *ACANTHOPTERYGII XIPHIIFORMES.*

Fam. **XIPHIIDÆ.**

51. **Xiphias gladius** *L.* **Swordfish.**
Casual visitant, of extremely rare occurrence.

> Filey, one, Sept., 1808, eleven feet long, twenty-three stones weight (Murray, 1832).
>
> Coatham, Redcar, one stranded in the winter, about 1874 (Nelson, MS.).
>
> Ulrome: formerly caught in the stake-nets here (Boynton, MS.).
>
> Bridlington Quay, occasionally (Boynton, MS.).
>
> Believed to have occurred also at Whitby and Scarborough (Meynell, 1844).

Div. *ACANTHOPTERYGII TRICHIURIFORMES.*

Fam. **TRICHIURIDÆ.**

52. **Lepidopus caudatus** (*Euphr.*). **Scabbard Fish.**

53. **Trichiurus lepturus** *L.* **Hairtail.**

Div. *ACANTHOPTERYGII COTTO-SCOMBRIFORMES.*

Fam. CARANGIDÆ.

54. Caranx trachurus (*L.*). Horse Mackerel. Scad.

Resident, common off the whole coast in summer, retiring to deeper water in winter.

55. Naucrates ductor (*L.*). Pilot-Fish.

56. Lichia glauca (*L.*). Derbio.

57. Capros aper (*L.*). Boar-Fish.

Accidental visitant from more southern seas, of extremely rare occurrence.

Redcar (Ferguson, Nat. Hist. of R., 1860).

Humber Mouth, one in 1877 (Cordeaux, Zool., 1879, p. 342).

Fam. CYTTIDÆ.

58. Zeus faber *L.* Doree. John Doree.

Casual visitant, occasionally taken at various places along the coast.

Fam. STROMATEIDÆ.

59. Centrolophus britannicus *Günth.*

60. Centrolophus pompilus (*L.*). Black-Fish.

Accidental visitant from southern seas, of extremely rare occurrence.

Redcar, one, Feb. 1852 (Rudd, Zool., 1852, p. 3504).

Fam. CORYPHÆNIDÆ.

61. Brama raii *Bl.* Ray's Sea-Bream.

Frequently cast ashore at Redcar during severe weather in the months of October, November, and December.

The species was originally described by Willughby and Ray from a specimen which had been left dead by the receding tide in Middlesburgh Marsh, at the mouth of the Tees, September 18th, 1681.

One was found in 1821 at Stockton-on-Tees (Yarrell and Day, *fide* Hogg).

Its appearance at Redcar—according to the observations of Messrs. T. S. Rudd and D. Ferguson, made from 1844 to 1852—is irregular, some years having been remarkable for their total absence, and others for their abundance; as many as twelve have been recorded as having occurred in one single morning.

At Bridlington Quay one was taken on the 1st of October, 1850, one on the 5th of November of the same year, and another on September 4th, 1851 (Boynton, MS.).

62. Lampris luna (*Gm.*). Opah. King-fish.

Casual visitant, of rare occurrence.

Filey Bay, one about 1767 (Pennant).

Whitby, one in 1807 (Hinderwell's History of Scarborough, 2nd ed., 1811).

Bridlington, one at the entrance to the harbour, 1809 (Id.).

Dogger Bank, one in 1838 (Yarrell).

Bridlington, one in 1842, weighing four stones and one pound (Meynell).

Bridlington, one, Sept., 1847 (Boynton, MS.).

Flamborough Head, one, Feb., 1849 (Norman, Zool., 1849, p. 2397).

Redcar, one, Nov., 1850 (Rudd, Zool., 1851, p. 3010).

Flamborough, one shot by Mr. M. Bailey, Oct., 1857 (Bailey, MS.).

Bridlington, one, Dec. 12, 1862 (Boynton, MS.).

Bridlington, one, Sept. 15, 1867 (Boynton, MS.).

Whitby, one caught in 1869, now in the Whitby Museum (Stephenson, MS.).

63. Luvarus imperialis *Rafin.*

Fam. **SCOMBRIDÆ.**

64. Scomber scomber *L.* Mackerel.

Appears off the Yorkshire coast from July to October, immense shoals in August and September.

65 Scomber colias *L.* Spanish Mackerel.

Accidental visitant from the Mediterranean Sea, of extremely rare occurrence.

Bridlington, one caught in 1861 by M. Walkington (T. Boynton, MS.).

66. Orcynus thynnus (*L.*). Tunny.

Accidental visitant from more southern seas, of extremely rare occurrence.

Bridlington, one, seven or eight feet long, 'a few years ago' (Meynell, 1844).

Tees mouth, one stranded, Sept., 1853, or October, 1854, said to be about 480 pounds weight (Hogg, Ann. & Mag. Nat. Hist., 1855, p. 213: Zool., 1855, p. 4594–6).

67. Orcynus germo (*Lacép.*). Germon.

68. Thynnus pelamys *C. & V.* Bonito.

69. Pelamys sarda (*Bl.*). Belted Bonito. Pelamid.

70. Auxis rochei (*Risso*). Plain Bonito.

71. Echeneis remora *L.* Remora.

Fam. **TRACHINIDÆ.**

72. Trachinus draco *L.* Great Weever.

Resident along the coast, but not very numerous. Known as 'Stingbull' or 'Cat-fish.'

73. Trachinus vipera *C. & V.* Lesser Weever.

Resident, abundant everywhere inshore. At Whitby it is called 'Natter,' and known on the Yorkshire coast generally as 'Sting-fish.'

Fam. **PEDICULATI**.

74. **Lophius piscatorius** *L.* **Angler.**

Resident, common along the coast. Sometimes attains to a very large size, specimens having been taken upwards of five feet in length.

Locally known on the coast as 'Devil,' 'Devil-fish,' or 'Pocket-fish'; small specimens are called 'Toad-fish.'

Fam. **COTTIDÆ**.

75. **Cottus gobio** *L.* **River Bullhead.**

Freshwater resident, generally distributed in streams below 900 feet altitude. Also in Malham Tarn, at an elevation of 1250 feet.

On the Lancashire border, near Clitheroe, this fish is known as 'Clot-head'; and about Dent it is called 'Bully-frog,' or more shortly 'Bully.'

76. **Cottus scorpius** *L.* **Short-spined Sea-Bullhead.**

Resident, abundant along the coast.

77. **Cottus bubalis** *Euphrasen.* **Long-spined Sea-Bull-head. Father-lasher.**

Resident, abundant along the coast.

78. **Cottus quadricornis** *L.* **Four-horned Bullhead.**

Has been reported to us as common in rock pools at Scarborough—a statement as to the correctness of which we entertain grave doubts.

79. **Trigla cuculus** *L.* **Red Gurnard.**

Resident along the coast, but not very common.

80. **Trigla lineata** *Gm.* **Streaked Gurnard.**

Reported to us as common at Scarborough, and occasional in Bridlington Bay.

81. **Trigla hirundo** *L.* **Sapphirine Gurnard. Tubfish.**

Occasionally taken off the Yorkshire coast. Appears to frequent deep water at a considerable distance from the shore, being usually brought in by fishing boats. Has once been stranded at the mouth of the Ouse below Swinefleet.

82. **Trigla gurnardus** *L.* **Grey Gurnard.**
Resident, abundant along the coast. The Redcar fishermen call it 'Groan.'

83. **Trigla lyra** *L.* **Piper.**
.Casual visitant, occasionally observed at Redcar, Whitby, and Scarborough. Locally known at Redcar as "Tom-piper.'

84. **Trigla obscura** *L.* **Shining Gurnard.**

Fam. CATAPHRACTI.

85. **Agonus cataphractus** (*L.*). **Pogge. Armed Bull-head.**
Resident, common along the coast. Specimens have been sent to us from Spurn, where they are commonly taken in lobster-pots, under the name of 'Rough-noses.'

86. **Peristethus cataphractus** (*L.*). **Mailed Gurnard.**

Div. *ACANTHOPTERYGII GOBIIFORMES.*

Fam. DISCOBOLI.

87. **Cyclopterus lumpus** *L.* **Lumpsucker.**
Resident, common along the coast. Very frequently stranded in boisterous weather. Known to the Flamborough fishermen as 'Redweens.'

88. **Liparis vulgaris** *Flem.* **Unctuous Sucker. Sea-Snail.**
Common, doubtless along the whole coast, though as yet only reported from Redcar, Scarborough, and Bridlington, at which places it is occasionally taken.

89. **Liparis montagui** (*Donov.*). **Montagu's Sucker.**

Fam. GOBIIDÆ.

90. **Gobius niger** *L.* **Black Goby. Rock Goby.**
Resident in rock pools along the coast, common at Redcar and Scarborough; probably equally so elsewhere, though not recorded.

91. **Gobius rhodopterus** *Günth.* Speckled Goby.

92. **Gobius paganellus** *L.*

93. **Gobius minutus** *Gm.* Spotted Goby. Little Goby.
Is taken at Scarborough. In 1844, Mr. Rudd recorded it as
being abundant in the pools on West Coatham Marshes,
at the Tees mouth (Zool., 1844, p. 395).

94. **Gobius ruthensparri** *Euphr.* Two-spotted Goby.
Resident, common in rock pools from Redcar to Flamborough.

Gobius gracilis *Jenyns.* Slender Goby.
Two specimens from the stomach of a cod taken at Redcar
agreed closely with this species (Rudd, Zool., 1844, p. 395).
Dr. Günther, however, in his Catalogue of Fishes, expresses
some doubt as to the validity of the species.

95. **Latrunculus albus** (*Parnell*). White Goby.

96. **Callionymus lyra** *L.* Dragonet. Skulpin.
Occasionally taken off the coast in deep water, but not
common. Dr. Günther unites the 'gemmeous' (♂) and
the 'sordid' (♀) dragonets as the two sexes of this species.

Div. *ACANTHOPTERYGII BLENNIIFORMES.*

Fam. **CEPOLIDÆ.**

97. **Cepola rubescens** *L.* Red Band-fish.
Mr. Cordeaux informs us that this species has been thrown
on the shore after storms.

Fam. **BLENNIIDÆ.**

98. **Anarrhichas lupus** *L.* Wolf-fish.
Resident, common along the coast. Called 'Wauffs' at
Redcar, 'Wuffs' at Whitby. The Yorkshire fishermen
esteem it highly as an article of diet, describing it as 'the
best fish that swims.'

99. **Blennius gattorugine** *Bl.* Gattoruginous Blenny.

100. **Blennius ocellaris** *L.* Ocellated Blenny. Butter-fly-fish.

101. **Blennius galerita** *L.* Montagu's Blenny.

102. **Blennius pholis** *L.* Shanny.

Resident, common in rock-pools from Redcar to Flamborough.

103. **Blenniops ascanii** (*Walb.*). Crested Blenny.

Resident, extremely rare. Has been included in the Scarborough list by Dr. Murray (1832) under the name of *'Blennius galerita,* Crested Blenny.' In the 2nd Edition of Yarrell's British Fishes, one is recorded as having been taken at Redcar in September, 1835, by Mr. T. P. Teale; it was a very fine specimen, measuring 6¾ inches in length.

This species is also described by authors under the name of 'Yarrell's Blenny.'

104. **Centronotus gunellus** (*L.*). Butterfish. Spotted Gunnel.

Resident, not uncommon, probably distributed along the whole coast, though as yet reported only from Redcar, Whitby and Scarborough.

105. **Zoarces viviparus** *L.* Viviparous Blenny.

Resident, common along the whole coast. This species is known as 'Burbot-Eel,' as we have ascertained by the examination of specimens sent to us from Spurn under that name.

Div. *ACANTHOPTERYGII MUGILIFORMES.*

Fam. **ATHERINIDÆ.**

106. **Atherina presbyter** *Cuv.* Sand-Smelt. Atherine.

Reported by Meynell in 1844 as taken in Bridlington Bay. It will be interesting to ascertain whether this species really

I

is taken on the Yorkshire coast, as from the researches of
Montagu and Yarrell it is apparently absent from the east
coast of England, its place being taken by the true Smelt
(*Osmerus eperlanus*), which, in its turn, is absent from the
south coast, where the Atherine is abundant.

107. **Atherina boyeri** *Risso.*   **Boyer's Atherine.**

Fam. **MUGILIDÆ.**

108. **Mugil octoradiatus** *Günth.*   **Eight-rayed Mullet.**

109. **Mugil capito** *Cuv.*   **Grey Mullet.**

Resident, local, not abundant, except at Spurn Point, where
Mr. Winson reports it as common in July.   The 'Grey
Mullet' is included in the lists for Scarborough (Murray,
1832) and for Redcar (Ferguson, 1860).   It is occasionally
caught in summer at Whitby; and has been taken at
Skinningrove, near Saltburn.

The distribution on the Yorkshire coast of the various species
of Grey Mullets as described by Dr. Günther—since the
publication of the last edition of Yarrell's B. Fishes—
requires investigation.   Dr. Günther informs us that he is
of opinion that all the species may possibly occur on the
east coast.

110. **Mugil auratus** *Risso.*   **Long-finned Grey Mullet.**

111. **Mugil septentrionalis** *Günth.*   **Lesser Grey Mullet.**

The only authority for including this species in the Yorkshire
list is the statement by Meynell (1844) that the two species
of Grey Mullet (*M. chelo* and *M. capito*) are occasionally
taken on the Yorkshire coast.

Div. *ACANTHOPTERYGII GASTROSTEIFORMES.*

Fam. **GASTROSTEIDÆ.**

112. **Gastrosteus aculeatus** *L.*   **Three-spined Stickleback.**

Resident, inhabiting fresh, salt, and brackish water indis
criminately, abundant and generally distributed.   In fresh

waters it inhabits canals, ditches, and sluggish streams, and also Malham Tarn at a height of 1250 feet above sea level. Known as 'Pricky' about Northallerton. The varieties, three of which may be expected to occur, have not been discriminated from the type, except by Mr. Thos. Boynton, of Ulrome, who finds the Smooth-tailed Stickleback (var. *gymnurus Cuv.*) and the Rough-tailed Stickleback (var. *trachurus C. & V.*) equally common.

113. **Gastrosteus brachycentrus** *C. & V.* **Short-spined Stickleback.**

114. **Gastrosteus spinulosus** *Jen.* **Four-spined Stickleback.**

115. **Gastrosteus pungitius** *L.* **Ten-spined Stickleback.**
Freshwater resident, locally distributed. Has been found near Leeds, Doncaster, Ulleskelfe, Thirsk, Slingsby, Redcar, and Ulrome.
This species is also known by the name of 'Tinker.'

116. **Gastrosteus spinachia** *L.* **Fifteen-spined Stickleback.**
Resident, not uncommon along the coast. Messrs. Cordeaux and Stephenson note that it frequents the 'mouths of harbours and docks.'

Div. *ACANTHOPTERYGII CENTRISCIFORMES.*

Fam. **CENTRISCIDÆ.**

117. **Centriscus scolopax** *L.* **Trumpet-fish. Sea-Snipe.**

Div. *ACANTHOPTERYGII GOBIESOCIFORMES.*

Fam. **GOBIESOCIDÆ.**

118. **Lepadogaster gouanii** *Lacép.* **Cornish Sucker.**

119. **Lepadogaster candollii** *Risso.* **Connemara Sucker.**

120. **Lepadogaster bimaculatus** *(Penn.).* **Network Sucker. Bimaculated Sucker.**

Div. *ACANTHOPTERYGII TÆNIIFORMES.*

Fam. **TRACHYPTERIDÆ.**

121. **Trachypterus arcticus** (*Brünn.*). **Deal-fish. Vaag-maer.**

122. **Regalecus banksii** (*C. & V.*). **Ribbon-fish. Banks' Oar-fish.**

An abyssal form, rarely cast ashore on the Yorkshire coast. To the five known Yorkshire instances, we are enabled, through the instrumentality of Mr. Thomas Stephenson, to add two which have hitherto remained unrecorded.

Whitby, one, 11⅓ feet long, Jan. 22, 1759 (Günther, Cat. Fishes, iii. 310). This appears to be the first recorded British specimen.

Filey, one, 13 feet long, March 18, 1796 (Günther, *fide* Banks, loc. cit.).

Yorkshire, 1845 (?), one, 24 (?) feet long (Günther, loc. cit.).

Redcar, one, 12 feet long, 1850 (Gray, P.Z.S., 1850, p. 52).

Whitby, one, 10 feet long, cast on the West sands, April 23, 1866; now in the Whitby Museum (Stephenson, MS.).

Whitby, one, 11 feet long, cast up on the beach east of the town, Oct. 2nd, 1870; now in the Whitby Museum (Stephenson, MS.).

Staithes, one, 16 feet long, end of January, 1880 (C. W. Elliott, Field, Feb. 7, 1880; E. W. H. Holdsworth, Field, Feb. 21, 1880). This example was left among the rocks by a retiring tide, and in its struggles to escape broke itself into three parts.

123. **Regalecus grillii** (*Lindr.*). **Sild-Kung.**

Ord. 2.
ACANTHOPTERYGII PHARYNGOGNATHI.

Fam. LABRIDÆ.

124. **Labrus maculatus** *Bl.* **Ballan Wrasse.**

Resident, abundant, frequenting the rocky coast from Redcar to Flamborough. In 1769, Pennant recorded it as appearing during summer in great shoals off Filey Brigg—a statement which has been copied by many subsequent writers. Donovan, in 1808, added that the Scarborough fishermen call it ' Old Wife.' Other names are given to it on the coast, it being known to the Flamborough fishermen as ' Old Ewe,' and at Redcar as ' Servellan wrasse ' or ' Sweet-lips.'

125. **Labrus mixtus** *L.* **Striped Wrasse. Cook Wrasse.**

Mr. Cordeaux remembers seeing an example some years ago on the Yorkshire coast, he believes at Flamborough (MS.).

126. **Crenilabrus melops** (*L.*). **Goldsinny. Corkwing.**

Mr. Cordeaux informs us that he thinks he has several times seen it in rock pools at Flamborough, though he never succeeded in capturing one.

127. **Ctenolabrus rupestris** (*L.*). **Jago's Goldsinny.**

Extremely rare ; has only occurred at Redcar, where but four specimens have been taken (Meynell, 1844). From one of these—sent by Mr. T. S. Rudd to Mr. Yarrell—the woodcut at p. 509 of the first volume of the 3rd Edition of the British Fishes was engraved.

128. **Acantholabrus palloni** (*Risso*).

129. **Centrolabrus exoletus** (*L.*). **Small-mouthed Wrasse.**

130. **Coris julis** (*L.*). **Rainbow Wrasse.**

131. **Coris giofredi** (*Risso*).

Ord. 3. **ANACANTHINI.**

Div. 1. *ANACANTHINI GADOIDEI.*

Fam. **LYCODIDÆ.**

132. Gymnelis imberbis (*L.*). Beardless Ophidium.

Fam. **GADIDÆ.**

133. Gadus morrhua *L.* Common Cod.

Resident, extremely abundant off the coast. Codlings are taken plentifully inshore, and ascend the estuary of the Humber as far as Goole.

134. Gadus æglefinus *L.* Haddock.

Resident, very abundant off the coast.

135. Gadus merlangus *L.* Whiting.

Resident, very abundant. Several young were netted in a warping-drain which communicates with the Ouse, at Goole, on the 30th of April, 1881, as Mr. Bunker informs us.

136. Gadus minutus *L.* Poor, or Power-Cod.

Reported as occurring on the rocky shores at Flamborough.

137. Gadus luscus *L.* Bib. Pout.

Resident, not uncommon off the coast. Yarrell gives this species the name of 'Whiting-pout.' Known to the Scarborough fish-dealers as 'John Doree'; and at Whitby it is another of the species called 'Old Wife.' Yarrell gives 'Kleg' as a name also in vogue at Scarborough.

138. Gadus poutassou *Risso.* Couch's Whiting.

139. Gadus pollachius *L.* Pollack.

Resident, common along the coast. Called 'Laits' at Whitby, 'Leets' at Scarborough.

140. **Gadus virens** *L.* **Coal-fish.**

Resident, abundant along the coast. This species, at various
stages of growth, receives divers names. The small fry, to
about nine inches in length, are called 'Pennocks,' and
afterwards become 'Billets' or 'Billards.' Other names in
use are 'Sillocks,' and at Spurn, 'Blue-backs.'

141. **Merluccius vulgaris** *Flem.* **Hake.**

Resident, seldom caught near the Yorkshire coast, but is
occasionally taken on the Dogger Bank, and becomes
more plentiful towards the coast of Holland.

142. **Phycis blennioides** (*Brünn.*). **Great Forked Beard.**

143. **Lota vulgaris** *Cuv.* **Burbot. Eelpout.**

Freshwater resident, local, occurring in sluggish rivers, but is
far from being a numerous species. Is reported as comparatively common in the river Hull, the Lower Derwent,
the Wiske, the Foss, the Ouse below Naburn, and in dykes
about Selby; and as scarce in the Seven, Pickering beck,
and other tributaries of the Upper Derwent, the Codbeck,
the Nidd, and the Wharfe.

144. **Molva vulgaris** *Flem.* **Ling.**

Resident, abundant off the coast. Mr. Yarrell states that in
Yorkshire the young are known as 'Drizzles.'

145. **Motella mustela** (*L.*). **Five-bearded Rockling.**

Resident, common on rocky shores from Redcar to Flamborough.

146. **Motella tricirrata** (*Bl.*). **Three-bearded Rockling.**

Resident, common on rocky shores from Redcar to Flamborough.

[The 'Mackerel Midges'—as the young of the various species
of Rocklings are called—are abundant in rock pools.]

147. **Motella maculata** (*Risso*). **Spotted Rockling.**

148. **Motella cimbria** (*L.*). **Four-bearded Rockling.**

149. **Motella macrophthalma** *Günther.*

150. **Raniceps trifurcus** (*Walb.*).   **Tadpole Hake.**

Extremely rare, has only occurred at Redcar.   In 1844,
Meynell stated that it was taken there; and in 1852, one
was observed by Mr. Rudd, in February or March (Zool.,
1852, p. 3504).

This is the Lesser Forked Beard, or Trifurcated Hake of
various authors.

151. **Brosmius brosme** (*Müll.*).   **Torsk.   Tusk.**

This species is included in Ferguson's Redcar list (1860);
has been caught off Whitby during the present season;
and Mr. H. Mudd, smack-owner, of Great Grimsby, informs
us that ' the Torsk is frequently caught near the coast.'

Fam. **OPHIDIIDÆ.**

152. **Ophidium broussonetii** *Müll.*

153. **Fierasfer dentatus** *Cuv.*   **Drummond's Fierasfer.**

154. **Ammodytes lanceolatus** *Lesauv.*   **Greater Sand-
Launce.**

Resident, common, but somewhat local, affecting sandy
shores.   This species is also known as Greater Sand-Eel,
and is the *Ammodytes tobianus* of Yarrell.

155. **Ammodytes tobianus** *L.*   **Lesser Sand-Launce.**

Resident, abundant everywhere on sandy shores.   Also
known as Lesser Sand-Eel.

156. **Ammodytes siculus** *Swains.*

Div. 2.   *ANACANTHINI PLEURONECTOIDEI.*

Fam. **PLEURONECTIDÆ.**

157. **Hippoglossus vulgaris** *Flem.*   **Holibut.**

Resident off the coast in moderate abundance.   Seldom
caught now at Flamborough, where thirty or forty years ago
they were very common.   Large specimens are reported to
us by some of our correspondents, one of which measured
fully six feet in length.

158. **Hippoglossoides limandoides** (*Bl.*). **Rough Dab. Sandsucker.**
Inhabits deep water in the North Sea, and is occasionally taken at Whitby.

159. **Rhombus maximus** (*L.*). **Turbot.**
Resident off the coast in moderate abundance.

160. **Rhombus lævis** (*L.*). **Brill.**
Resident off the coast, not uncommon.

161. **Rhombus megastoma** (*Donov.*). **Sail-Fluke. Whiff.**
Resident in moderate numbers.

162. **Rhombus norwegicus** *Günth.* **Ekstrom's Topknot.**

163. **Rhombus punctatus** *Bl.* **Müller's Topknot.**
Of very rare occurrence.

Redcar, several specimens were found on the beach in 1836, none since (Meynell, 1844).
Redcar, two in February, 1852 (Rudd, Zool., 1852, p. 3504).
Redcar, one in April, 1852 (Ferguson, Nat., 1852, p. 134).
Bridlington, one taken May 27, 1881 (Boynton, MS.).

164. **Phrynorhombus unimaculatus** (*Risso*). **Bloch's Topknot.**
Is said to have occurred at Redcar, and is included in Ferguson's list for that place (1860).

165. **Arnoglossus laterna** (*Walb.*). **Scaldfish. Megrim.**

166. **Pleuronectes platessa** *L.* **Plaice.**
Resident, very common, but Mr. H. Mudd informs us that it is becoming more scarce, and that it is found in greatest numbers on or near the Dogger Bank.

167. **Pleuronectes limanda** *L.* **Common Dab.**
Resident, very abundant. Known as 'Sand-dab' at Redcar.

168. **Pleuronectes microcephalus** *Donov.* **Smear Dab.**
Resident, common.

This is the 'Smooth Dab' or 'Lemon Dab' of Yarrell.

169. **Pleuronectes cynoglossus** *L.* **Craig-Fluke. Pole.**

170. **Pleuronectes elongatus** (*Yarrell*). **Long Flounder.**

171. **Pleuronectes flesus** *L.* **Flounder.**
Resident, abundant along the coast and in rivers within and
even beyond the influence of the tide. The distance to
which it now attains commonly in the rivers is consider-
ably curtailed by the comparatively recent erection of
numerous dams and weirs. A few, however, small in size,
are found in the Nidd at Wilstrop and Kirk Hammerton.
Ascends the river Ribble as far as Mytton every year, in
May or June, leaving again in September.

172. **Solea vulgaris** *Quensel.* **Sole.**
Resident, common off the coast. Has once occurred at
Goole, in the estuary of the Humber, near the mouth of
the Ouse.

173. **Solea aurantiaca** *Günth.* **Lemon Sole.**
Resident, not uncommon.

174. **Solea variegata** (*Donov.*). **Variegated Sole.**

175. **Solea minuta** (*Parnell*). **Dwarf Sole.**
Reported as having been taken at Whitby.

This species is also called 'Solenette' or 'Little Sole' by
authors.

Order 4. **PHYSOSTOMI.**

Fam. **CYPRINIDÆ.**

Sub-fam. *CYPRININA.*

176. **Cyprinus carpio** *L.* **Common Carp.**
Introduced. Occurs in the Swale, Codbeck, and Grimescar
Beck, and also in artificial waters throughout the county.

177. **Carassius vulgaris** *Nilss.* **Crucian Carp.**

This species and its variety, **C. gibelio** *Bl.*, or **Prussian Carp,** have been introduced into numerous ponds.

178. **Carassius auratus** (*L.*). **Gold Carp.**

Introduced into ponds and reservoirs.

179. **Barbus vulgaris** *Flem.* **Barbel.**

Freshwater resident, locally distributed in the rivers of the central plain, except the Derwent and the polluted waters of the Aire, Calder, Dearne, and Don.

180. **Gobio fluviatilis** *Flem.* **Gudgeon.**

Freshwater resident, common and generally distributed in rivers and ponds.

Sub-fam. *LEUCISCINA.*

181. **Leuciscus rutilus** (*L.*). **Roach.**

Freshwater resident, abundant and generally distributed, except in the North-western Fell district and the polluted rivers, from which it is absent. Has been introduced into canals, reservoirs, and ponds.

182. **Leuciscus cephalus** (*L.*). **Chub.**

Freshwater resident, abundant and generally distributed. Absent from the rivers of the north-west, and from the polluted portions of those of the manufacturing districts.

183. **Leuciscus vulgaris** *Flem.* **Dace.**

Freshwater resident, generally distributed and common, being absent only from polluted rivers and hill-streams.

184. **Leuciscus erythrophthalmus** (*L.*). **Rudd.**

Freshwater resident, abundant in Holderness, where it occurs in ponds. Reported as in Lake Semerwater, and ponds in other parts of the county, and also in the Codbeck. A few have been caught in the Ouse at York.

185. Leuciscus phoxinus (*L.*)　Minnow.

Freshwater resident, generally distributed and very abundant. Occurs in the streams of the Fell district up to an elevation of 1000 feet, and in Malham Tarn at 1250 feet.

186. Tinca vulgaris *Cuv.*　Tench.

Freshwater resident, introduced into numerous ponds, reservoirs and canals throughout the county, and into some sluggish streams.

Sub-fam. *ABRAMIDINA.*

187. Abramis brama (*L.*).　Bream.

Freshwater resident, local, found only in the lower reaches of the Don, Wharfe, Ure, Codbeck, Derwent, and Hull, and in canals. The commonest fish of Lake Semerwater, where it has probably been introduced.

188. Abramis blicca (*Bl.*).　White Bream.　Breamflat.

Freshwater resident, very local. Is common about Goole, and in the Foss and Ouse near York, and is occasionally taken in Holderness.

189. Alburnus lucidus *Heck. & Kner.*　Bleak.

Freshwater resident, occurring commonly in the lower waters of the Tees, Leven, and of the rivers of the Central Plain, and also in the Hull.

Sub-fam. *COBITIDINA.*

190. Nemachilus barbatulus (*L.*).　Loach.

Freshwater resident, generally distributed and abundant in shallow waters. Found in the streams of the Northwestern Fells up to 900 feet, and also in Malham Tarn (1250 feet). Absent from the Goole district.

191. Cobitis tænia *L.*　Spinous Loach.

Fam. **SCOMBRESOCIDÆ.**

192. Belone vulgaris *Flem.*　Gar-fish.

Resident off the coast; not uncommon; approaching the shore in summer; in autumn it has been cast ashore in considerable numbers during heavy gales. Known at Redcar as 'Long-nose.'

193. Scombresox saurus (*Walb.*). Saury. Skipper.

Occasionally taken in harbours and shallow water. Reported from Scarborough only.

194. Exocœtus evolans *L.* Flying-fish.

195. Exocœtus volitans *L.* Greater Flying-fish.

Fam. **ESOCIDÆ.**

196. Esox lucius *L.* Pike.

Freshwater resident, abundant in the 'deeps' of the middle and lower reaches of all the rivers except the polluted ones ; also in ponds, canals, and reservoirs, into which it has been introduced.

Fam. **STERNOPTYCHIDÆ.**

197. Maurolicus borealis (*Nilss.*). Argentine. Pearlside.

Of rare occurrence at Redcar, where it has been found from January to May. Mr. T. S. Rudd first met with it there in May, 1841 ; in 1843, he took thirteen specimens; in Feb., 1851, four; and in February and March, 1852, forty. Elsewhere in the British Isles it appears only to have occurred singly. The species is represented in the British Museum by Yorkshire specimens sent by Mr. Rudd.

Fam. **SALMONIDÆ.**

198. Salmo salar *L.* Salmon.

Freshwater resident, periodically descending to the sea, abounding in all the Yorkshire rivers except the Aire, from which it is excluded by pollution, and the Hull, Don, Wiske, and Codbeck, which are too sluggish. Into the Esk it was introduced some twelve years ago. The spawning season in the Yorkshire rivers is in November and December, immediately after which, in January and February, the kelts or spent-fish descend to the sea. The date of their return, as fresh-run fish, varies greatly, and is dependent upon the state of the rivers—if either July,

August, or September are wet, the Salmon commence
running from the sea—if otherwise, their ascent is delayed
until the autumn rains set in.

The eggs are usually deposited in November and
December, and take from ninety to one hundred and
thirty days to hatch, but this is greatly dependent on
the temperature of the water and other conditions. It
was formerly a prevalent belief that the fry, or 'parr' as
they are termed, migrated during the first year, a theory
not now entertained. The young fish, hatched say in
March, remain in the river until April or May of the year
following, that is for thirteen or fifteen months after leaving
the ova. They have by that time assumed the migratory
dress, thus becoming 'smolts,' and are carried down to
the sea with the first floods of May. Mr. Phillips considers that the weight of evidence favours the opinion that
smolts do not return the same year in which they enter the
sea, but that they remain there about fifteen months, and
ascend the river as 'grilse' along with the Salmon about
the month of August in the second season.

For much interesting information on Yorkshire Salmon
we are indebted to Mr. John H. Phillips, Secretary to the
Yorkshire Salmon Fishery Board, to Mr. Samuel Wilkin-
son, Secretary to the Esk Salmon Fishery Board, and to
Mr. H. T. Gardiner, of Goole.

199. **Salmo trutta** *Flem.* **Salmon-Trout. Sea-Trout.**

Freshwater resident, periodically descending to the sea,
present in all the rivers frequented by *S. Salar*, especially
in those flowing into the North Sea, and is particularly
abundant in the Esk and Tees, where it is known both as
'salmon-trout' and 'bull-trout.' Very abundant along the
coast, and was taken in numbers in the stake nets which
formerly existed in Bridlington Bay. The dates for this
species are substantially the same as those for the Salmon,
the spawning season (as given for the Esk) being in
November and December, the kelts descending during the
months of March, April, and May, and re-ascending in
June, the smolts descending in April and May.

The '**Bull-Trout**'—thoroughly and generally believed
in as a distinct species in Yorkshire—is very abundant in
the Esk and the Tees. Dr. Günther, however, states that
all specimens of the so-called 'Bull-Trout' that he has
examined have been referable to *S. trutta* (Salmon-trout),

S. cambricus (Sewin), or *S. fario* (Common Trout). An example of *S. trutta* from the Ouse is in the British Museum, and Dr. Günther has had specimens submitted to him from the Esk.

200. **Salmo cambricus** *Donov.* **Sewin.**

201. **Salmo fario** *L.* **Common Trout.**

Freshwater resident, generally distributed and abundant above an elevation of about 250 feet. Below this its numbers decrease, and the species becomes local, being found in sharp-running streams. As to the quality of Yorkshire trout there is considerable variation, even in those of contiguous streams presenting no marked physical differences. This is exemplified in the case of the rivers Aire, Wharfe, and Ribble, which rise in the same district, and all of which receive the drainage of the limestone plateau on which Malham Tarn is situate. The Tarn trout are reputed to be the best in flavour, and only inferior to the trout of Loch Leven. The Aire trout are almost as good, the Ribble trout being ranked next, while those of the Wharfe are considered inferior to them —the cause of the variation in quality being obscure, for there does not seem to be any marked difference in the nature of the river-beds. The Malham ova are much in request for stocking and improving other streams, and have been introduced into the Thames and the Wharfe, but with what effect has not been fully ascertained.

Remarkable malformations are observed in the trout of Malham Tarn, and of a beck on the western side of Penyghent. This is manifested in the former by the deficiency of the gill-cover in about one in every fifteen fish caught— a calculation based upon a statement with which Mr. Walter Morrison has furnished us of the total number caught from 1865 to 1880. In the case of the 'ground trout' of Penyghent, as they are called, Mr. John Foster informs us that the malformation consists of a singular projection of the under jaw beyond the upper. These aberrations are considered to be the result of interbreeding, due to an extreme degree of isolation. The isolation of Malham Tarn is complete; it has no feeders of sufficient size for the introduction of new blood, while the overflow is absorbed by fissures in the limestone, after being swallowed up by which the water re-appears—as the river

Aire—after a subterranean course of two miles. The beck at Penyghent is exceedingly small, and after a short half-mile course disappears in a similar manner.

In Malham Tarn there are two varieties of trout, the 'silver trout' and the 'yellow trout,' both of which are subject to the malformation before spoken of. The yellow trout were by Mr. Couch erroneously considered to be small examples of the Great Lake Trout (*Salmo ferox*), but on specimens of both forms being submitted to Dr. Günther they were pronounced to be of the same species.

202. **Salmo argenteus** (*C. & V.*). **Silvery Salmon.**

203. **Salmo brachypoma** *Günth.* **Short-headed Salmon.**

This species—one of the most easily recognisable of the salmonoids—was described by Dr. Günther from specimens from the Yorkshire Ouse, and from the Tweed and Forth, and is a migratory form. The Ouse specimens in the British Museum were presented in 1865 by Mr. J. H. Phillips, of the 'Yorkshire Salmon Fishery Preservation Society, Beadlam Grange, Nawton.'

204. **Salmo gallivensis** *Günth.* **Galway Sea-Trout.**

205. **Salmo orcadensis** *Günth.* **Loch-Stennis Trout.**

206. **Salmo ferox** *Jard. & Selby.* **Great Lake-Trout.**

207. **Salmo stomachicus** *Günth.* **Gillaroo Trout.**

208. **Salmo nigripinnis** *Günth.* **Black-finned Trout.**

209. **Salmo levenensis** *Walk.* **Loch-Leven Trout.**

210. **Salmo alpinus** *L.* **Alpine Charr.**

211. **Salmo killinensis** *Günth.* **Loch-Killin Charr.**

212. **Salmo willughbii** *Günth.* **Windermere Charr.**

213. **Salmo perisii** *Günth.* **Torgoch. Welsh Charr.**

214. **Salmo grayi** *Günth.* **Gray's Charr.**

215. **Salmo colii** *Günth.* **Cole's Charr.**

216. **Osmerus eperlanus** (*L.*). **Smelt.**

Resident and common in the estuaries of the Tees and Humber. It abounds in the Ouse and the Humber from Naburn Lock to Spurn Point, and occurs a considerable distance up the Tees. Yarrell states that Col. Meynell, of Yarm, kept Smelts for four years in a freshwater pond having no communication with the sea, where they continued to thrive and propagated abundantly. When the pond was drawn, the fishermen of the Tees considered that they had never seen a finer lot of Smelts; there being no loss of flavour or quality.

217. **Coregonus clupeoides** *Lacèp.* **Gwyniad. Powan.**

218. **Coregonus vandesius** *Rich.* **Vendace.**

219. **Coregonus pollan** *Thomp.* **Pollan.**

220. **Thymallus vulgaris** *Nilss.* **Grayling.**

Freshwater resident, found in varying abundance in the middle waters of the Wharfe, Washburn, Nidd, Ure, and Swale, also in the Cover, Wiske, and Codbeck, the Rye and other tributaries of the Upper Derwent, and in the Scalby Beck near Scarborough. In the Tees it is very limited in its numbers, and it has been introduced into the Esk. Formerly abundant in the Ribble and Hodder, their extreme scarcity—if not extinction—being ascribed to the great increase of Salmon. It was also formerly abundant in the Aire about Bingley, but in 1824 all were destroyed by the bursting of a peat bog; subsequent attempts at re-introduction have as yet proved ineffectual.

221. **Argentina silus** (*Ascanius*).

222. **Argentina sphyræna** *L.* **Hebridal Argentine.**

Extremely rare, there being only one record for the county.

Redcar, one found February 5, 1852, and submitted to Mr. Yarrell (Rudd, Zool., 1852, p. 3504).

Fam. **CLUPEIDÆ.**

223. **Engraulis encrasicholus** (*L.*). **Anchovy.**

K

224. Clupea harengus *L.* Herring.

Periodical visitant, occurring annually in shoals on the York-shire coast from July to November.   The Bay Fishery—as it is called at Bridlington and Flamborough—continues up to Christmas.

At Redcar a single example of the variety *Leachii* of Yarrell was found on the beach in April, 1843 (Rudd, Zool., 1844, p. 395).

225. Clupea sprattus *L.* Sprat.

Abundant on the coast; occasionally enters the Humber.

226. Clupea alosa *L.* Allis Shad.

Resident, not uncommon on the coast and in the Humber; is occasionally taken in the Ouse at Naburn.

227. Clupea finta *Cuv.* Twaite Shad.

Resident, not uncommon at Whitby and Bridlington.

This and the preceding species appear to be much con-founded with each other.

228. Clupea pilchardus *Walb.* Pilchard.

Casual visitant in summer, occasionally in some numbers.

Fam. MURÆNIDÆ.

229. Anguilla vulgaris *Flem.* Sharp-nosed Eel.

Freshwater resident, universally distributed and very abun-dant; periodically descending to the sea.

The range of this species is more extensive than that of any other fish; it is found in every part of Yorkshire, and occurs at an altitude of 1300 feet or more.

230. Anguilla latirostris *Risso.* Broad-nosed Eel.

Freshwater and marine resident, generally distributed and abundant, but not equally so with *A. acutirostris*, and appears to be confined to the deeper and lower parts of the rivers. This species is also known as 'Grig.'

231. Conger vulgaris *Cuv.* Conger.

Resident, abundant along the coast. Has occasionally been found washed ashore after severe gales. The young are occasionally netted at Goole.

Leptocephalus Morrisii *Gm.*, **Anglesea Morris,**—a specimen of which has been recorded as occurring at Redcar (Rudd, Zool., 1852, p. 3504)—is considered by Dr. Günther to be a larval form, and probably of the conger.

232. **Muræna helena** *L.* **Murry.**

Order 5. **LOPHOBRANCHII.**

Fam. **SYNGNATHIDÆ.**

233. **Siphonostoma typhle** (*L.*). **Deep-nosed Pipe-fish.**
Reported by Mr. J. W. Woodall as having occurred at Scarborough.
This species is the Broad-nosed Pipe-fish of Couch.

234. **Syngnathus acus** *L.* **Great Pipe-fish. Tangle-fish. Needle-fish.**
Resident, common along the coast.

235. **Syngnathus brevicaudatus** *Cornish.* **Blunt-tailed Pipe-fish.**

236. **Nerophis æquoreus** (*L.*). **Æquoreal Pipe-fish.**
One is recorded by Yarrell as having occurred at Scarborough, and a specimen in the Whitby Museum is believed by Mr. Martin Simpson to have been taken near that place.
This species is the 'Ocean Pipe-fish' of Couch.

237. **Nerophis ophidion** (*L.*). **Straight-nosed Pipe-fish.**
At Whitby in 1880 a specimen was taken in a lobster-pot (Stephenson, MS.). It has a place in Dr. Murray's Scarborough list (1832), and in Ferguson's Redcar list (1860).
This species is the 'Snake Pipe-fish' of Couch.

238. **Nerophis lumbriciformis** (*L.*). **Little Pipe-fish. Worm Pipe-fish.**

239. **Hippocampus antiquorum** *Leach.* **Sea-Horse.**
Accidental visitant of rare occurrence. A southern pelagic form, which has occurred three times at Whitby, and once at Bridlington.

Order 6. **PLECTOGNATHI.**

Fam. **SCLERODERMI.**

240. **Balistes maculatus** *Gm.*

241. **Balistes capriscus** *Gm.* **Pig-faced Trigger-fish. File-fish.**

Fam. **GYMNODONTES.**

242. **Tetrodon lagocephalus** *L.* **Crop-fish, or Pennant's Globe-fish.**

243. **Orthagoriscus mola** (*L.*). **Short Sun-fish.**
Casual visitant of rare occurrence.

Scarborough, an adult cast ashore weighing 120 lbs., now in the Museum, figured by Yarrell (Ed. 3, vol. ii., p. 432).
Redcar, one (Meynell, 1844).
Bridlington, two or three (Meynell, 1844).
Bridlington, one, Sept. 5th, 1840 (Boynton, MS.).
Bridlington, one, Sept., 1855 (Boynton, MS.).
Bridlington Bay, one caught, August, 1880 (Boynton, MS.).

244. **Orthagoriscus truncatus** (*Retz.*). **Oblong Sun-fish.**

Sub-class 3. *CYCLOSTOMATA.*

Fam. **PETROMYZONTIDÆ.**

245. **Petromyzon marinus** *L.* **Sea-Lamprey.**
Resident, not uncommon on the coast; has been caught in the Ouse.

246. **Petromyzon fluviatilis** *L.* **Lampern. River-Lamprey.**
Freshwater resident, generally distributed and abundant. Some of our correspondents allude to this species as being migratory, and in one or two instances give dates of arrival and departure. The Dutch fishermen have for more than a century visited the Ouse for the purpose of purchasing these fish for bait.

247. **Petromyzon branchialis** *L.* **Pride. Sandpiper. Small Lamprey.**

Freshwater resident, common in various localities. Probably more generally distributed, but much overlooked, being often confounded with the Lampern.

Fam. **MYXINIDÆ.**

248. **Myxine glutinosa** *L.* **Glutinous Hag. Borer.**

Resident, abundant off the whole coast from Redcar to Flamborough.

Sub-class 4. *LEPTOCARDII.*

Fam. **CIRROSTOMI.**

249. **Branchiostoma lanceolata** (*Pall.*). **Lancelet.**

Redcar (Yarrell, ed. iii., 1859, vol. i., p. 4).

APPENDIX.

MAMMALIA.

23. Martes sylvestris *Niss.* Marten (p. 6).

One shot in the woods, Azerley, near Ripon; now in the possession of Col. Crompton (Joseph Lucas, MS.).

One seen by Mr. J. Lucas, in 1870, on High Ash Head Moor, above Masham (MS.).

30. Phoca vitulina *L.* Common Seal (p. 8).

Respecting the date of extinction, information received from Mr. H. G. Faber, of Stockton-on-Tees, and Mr. H. T. Mennell, shows that the animal survived much later than is stated at p. 8.

The latter has furnished a copy of his and Mr. Perkins' list of the mammalia of Northumberland and Durham, published in 1863, wherein is stated that between 1820 and 1830 about a thousand seals frequented the mouth of the Tees, of which as many as thirty might often be counted at one time; but that in 1862 the number was reduced to three individuals. The seals exhibited great dread of the steamboats, which had greatly increased in number on the river during the preceding few years; and to this and the enormous increase of population in the neighbourhood, was attributed their rapid decrease.

Mr. Faber remembers the seals being numerous at the Tees mouth, and has seen them lying on the sands as many as a dozen together. He recollects disturbing one about twenty years ago on Seaton Snook, which was very tame, swimming about within twenty yards, and could only

be driven away by throwing stones at it. He adds that
the last 'native seal' was killed about ten years ago, when
it was shot from the Fifth Buoy Light. A small seal got
about a year ago was but a mere casual visitant. It will
thus be seen that the animal survived a good deal later
than the evidence available at the time of writing the list
would lead one to suppose.

BIRDS.

73. Lanius collurio *L.*   Red-backed Shrike (p. 27).
Nested at Swillington, near Leeds, in 1881 (J. Tennant, MS.).

133. Cypselus melba (*L.*).   White-bellied Swift (p. 37).
The date of the Hornsea occurrence is June 2nd, 1870
(Boyes, MS.).

203. Plegadis falcinellus (*L.*).   Glossy Ibis (p. 52).
Filey, an immature specimen obtained in 1863 (Roberts'
Scarborough list).

272. Otis tarda *L.*   Great Bustard (p. 65).
The only reference of early date to this bird is that the Earl
of Northumberland's regulations in 1512 for his castles of
'Wresill and Lekinfield in Yorkshire' included among the
articles for principal feasts 'Item Bustardes for my Lords
owne Meas at Principall Feists Ande noon outher tyme
Except my Lords comaundement be otherwis;' but no
price is attached as in the case of other birds mentioned.
Prof. Newton, of Magdalen College, Cambridge, kindly
communicates the following additional evidence :—
'Rather more than a year ago the Master of Trinity
College, Dr. W. H. Thompson, told me that when he was
about six or seven years old he was living at York with his
grandfather, to whom a Bustard was sent as a present.
Dr. Thompson remembered going into the servants' hall
or the kitchen to look at it, and some one was holding it
up by the legs. He thought it weighed about eight or
nine pounds, and it would therefore be a hen bird. He
supposed it had been shot on the Yorkshire Wolds. where

he had heard Bustards once existed, and that it was eaten in the house, but he had no recollection of having tasted it, or indeed anything more about it. Dr. Thompson graduated B.A. in 1832, and supposing him to have been then twenty-two years of age, the event must have happened about 1816 or 1817.

'In October, 1854, Mr. Barnard Henry Foord, of Foxholes, near Scarborough, then aged twenty-five, told me he remembered having seen Bustards—the last was at Foxholes about nineteen years before (*i.e.*, 1835). His father once saw eleven together. He had heard his uncle speak of running Bustards with greyhounds, as if he had been present at the time.

'This Mr. Foord is, I believe, now dead. I was very much struck at the time by the nature of his evidence, for I had believed that the bird was extinct in Yorkshire before 1835—and I remember pressing him particularly with questions on this point; but he persisted in the truth of his statement. I confess I was not, nor am I now, satisfied with it, though I am unable to suggest any explanation of the difficulty—for even if he had been a year or two older than he said (and he could not have been more), it would still remain.'

288. **Himantopus candidus** *Bonnat.* **Black-winged Stilt** (p. 73).

Miss Hall, of Scorborough, has informed Mr. Stephenson that the specimens now in his possession were shot about thirty years ago in Aike Carr, by Lord Hotham's keeper (John Stephenson, MS.).

289. **Phalaropus hyperboreus** (*L.*). **Red-necked Phalarope** (p. 73).

Tees mouth, four in 1854 (Rev. H. Smith, MS.).

ERRATUM.

Page 6, line 4 from the bottom, For **'PINE'** read **'BEECH.'**

AUTHORITIES CONSULTED.

Of **Published Records** the authors believe that comparatively little has escaped their notice, all available sources of information, whether natural history journals and publications, or the
appendices to topographical works, having been carefully
examined.

The PRINCIPAL DISTRICT-LISTS which have been contributed
to the natural history periodicals include Leyland's list of Halifax
birds (Loudon's Mag. N.H., 1828), Williamson's note on Scarborough birds (Proc. Zool. Soc., 1836), Denny's list of animals
occurring near Leeds (Ann. & Mag. N.H., 1840), abstracts of
Allis's report on the birds, and Meynell's on the fishes of Yorkshire (Rep. Brit. Ass., 1844), Hogg's 'Catalogue of Birds observed
in South-eastern Durham and in North-western Cleveland' (Zool.,
1845), and Talbot's Birds of Wakefield (Nat., 1876).

The chief lists which have appeared in or as appendices
to topographical works are to be found in Miller's History
of Doncaster (1804), Graves' History of Cleveland (1808),
Young's History of Whitby (1817), Whitaker's History of Richmondshire (1823), Hinderwell's History of Scarborough (1832),
Lankester's Account of Askern (1842), Barker's Three Days of
Wensleydale (1854), Hobkirk's History and Natural History of
Huddersfield (1859, second edition in 1868), Ferguson's Natural
History of Redcar (1860), Theakston's Scarborough Guide (1863
and subsequent editions), and Hobson's Life of Charles Waterton
(1866).

In addition to these lists, there are innumerable records in the
periodicals and in natural history and topographical works generally, chiefly in the pages of that well-known repertory of information, the 'Zoologist,' and also in the Ibis, the Field, Land
and Water, Loudon's Magazine of Natural History, the

various publications bearing the name of 'Naturalist' which have appeared under the editorship of Messrs. Wood, Morris, Hobkirk, Porritt, and others, and in the earlier volumes of the Annals and Magazine of Natural History, and of the Proceedings of the Zoological Society.

The most important and probably the only separate work devoted to any part of the Yorkshire vertebrate fauna is Mr. Cordeaux's Birds of the Humber District (1872), a valuable contribution—not only to the Yorkshire and Lincolnshire fauna—but to British ornithology generally, and abounding in original observations on the important subject of migration. For the whole district—the Yorkshire half of which is equivalent to Holderness and the Wolds—280 species are enumerated.

Of **Manuscript information,** the authors have had (through the kindness of his son-in-law, Mr. W. Pumphrey) the advantage of consulting the original MS. of Mr. Thomas Allis's Report on the Birds of Yorkshire (1844). This list, which includes 258 species, has been of much assistance from the evidence it contains as to the faunistic position of various species at that and earlier dates, besides the numerous records of occurrences.

Manuscript lists have been furnished by numerous naturalists of the county. Of these the following have contributed lists of the mammalia, reptiles, and freshwater fishes of their respective districts (those marked * having also contributed lists of birds):—

Messrs. *G. Abbey, jun. (Grinkle Park, near Whitby); *H. Andrews (Aldborough); W. Ingram Baynes (Ulleskelf); Geo. Bishop (Skipton); *Fred Boyes (Holderness); *Thomas Boynton (Bridlington); *James Brigham (Slingsby); W. B. Brigham (Driffield); C. J. E. Broughton (Wharncliffe); Thomas Bunker (Goole); Jno. Braim and R. Clarke (Pickering); John T. Calvert (Keighley); *James Carter (Masham); J. H. Carter (Spurn); J. W. Carter (Bradford); *T. R. Clapham (Austwick); Rev. E. Maule Cole, M.A. (Wetwang and Wolds); *John Cordeaux (Humber and Coast); *Capt. Wade-Dalton (Hawxwell); J. S. Davidson (Dent); H. W. T. Ellis (Thorne Waste); *John Emmet (Boston Spa); *John Foster (Upper Ribblesdale); Matthew Foster (Sancton); *John Grange (Harrogate); Rev. Geo. Hales (Barningham, Wycliffe, &c.); C. C. Hanson (Elland and Greetland); *John Harrison (Wilstrop); John T. Hausell (Thirsk); W. H. Hay (Scarcroft); *James Ingleby (Eavestone); Thomas Lister (Barnsley); William Lister (Glaisdale);

R. Lofthouse (Middlesborough); R. Morton Middleton, jun., F.L.S. (North-allerton, &c.); J. E. Miller (Middleham); *W. J. Milligan (Wetherby); *F. S. Mitchell (Mitton and Lower Ribblesdale); Walter Morrison (Malham Tarn); T. H. Nelson (Redcar and Teesmouth); George Page (Guisborough); F. Parkinson (Market Weighton); J. Petyt (Bolton Abbey); G. T. Porritt, F L.S. (Huddersfield); Walter Raine (Ryther, Leeds, &c.); F. G. S. Rawson (Halifax); Lord Ribblesdale (Gisburn); *J. H. Salter (Ackworth and Scarborough); William Scoby (Kirby Moorside); *Rev. H. H. Slater, B.A., F.Z.S. (Ripon); *Walter Stamper (Nunnington); *Thos. Stephenson (Whitby); *John Tennant (Wilstrop and Skewkirk); Richard G. Tuke (Castle Howard); J. A. Wheldon (Northallerton); and J. E. Clark. B.A.,B.Sc.,F.G.S., R. M. Christy, E. J. Gibbins, C. Helstrip, Wilson, and Woods (York).

Other contributors of bird-lists, in addition to those marked above with the asterisk, are :—

Messrs. Matthew Bailey (Flamborough); W. W. Boulton (Beverley); E. P. P. Butterfield (Wilsden); Rev. J. W. Chaloner (Newton Kyme); R. B. Cragg (Skipton); Charles Dixon (Sheffield); Alfred Jackson, M.D. (Market Weighton); Major Lawson (Bridlington); Robert Lee (Thirsk); R. Richardson (Beverley); Alfred Roberts (Scarborough); Rev. Henry Smith (Redcar); Edward Tindall (Knapton); Capt. E. H. Turton (Upsall Castle); J. W. Davis (Halifax); and James Varley (Huddersfield).

Lists of marine fishes have been contributed by the following:—

Messrs. Matthew Bailey (Flamborough); Thomas Boynton (Bridlington Bay); John Cordeaux (Holderness Coast); T. H. Nelson (Redcar); Thomas Stephenson (Whitby); T. Winson (Spurn Point); and J. W. Woodall (Scarborough).

To all these contributors of local lists the best thanks of the authors are due, for it is they who—by furnishing the full and indispensable details—have made it possible to give with an approximation to accuracy the actual geographical range of the various species. Hence it will be seen that the few words allotted to each species are founded upon a broad and substantial substratum of facts contributed by numerous observers, and that it is only in the few instances where records are vouched for by single observers that the citation of authorities has been possible in the text.

But the contributors of lists are not the only persons to whom the authors are indebted for assistance. Acknowledgment is due to Professor Alfred Newton, M.A., F.R S., Dr. Günther, F.R.S.,

Mr. A. G. More, F.L.S., Dr. G. E. Dobson, F.L.S., Mr. H. E. Dresser, F.L.S., Mr. J. E. Harting, F.L.S., and above all to Mr. John Cordeaux and Mr. Thomas Stephenson, for the very kind manner in which numerous queries were replied to, while for scattered facts and information upon special points the

Rev. J. C. Atkinson, B.A.; Mr. F. Bond, F.Z.S.; Mr. J. D. Butterell; Rev. H. C. Casson; Hon. Payan Dawnay; Hon. F. H. Dawnay; Mr. N. F. Dobrée; Mr. George Edson; Mr. H. G. Faber; Dr. Francis M. Foster; Mr. John Grassham; Mr. J. H. Gurney, jun., F.Z.S.; Rev. F. W. Hayden; Mr. H. B. Hewetson; Mr. Francis Hoare; Mr. Henry Kerr; Mr. P. W. Lawton; Mr. H. T. Mennell, F.L.S.; Mr. Thomas Machen; Prof. L. C. Miall, F.G.S.; Mr. C. C. Oxley; Rev. G. E. Park; Mr. Thomas Southwell, F.Z.S.; Mr. John Stephenson; Mr. W. H. Taylor; and Mr. J. Whitaker, F.Z.S.,

besides some of the contributors of lists, have placed the authors under obligation.

Although the limited space which can be allotted to the subject in the present work precludes a fuller use of the mass of valuable information which the courtesy of correspondents has placed at their disposal, the authors hope to have the pleasure of utilising it in future publications, and for such purpose would be glad to receive additional notes. Mr. Roebuck would be pleased to receive records of the occurrence of the mammalia and reptiles, more particularly of the smaller and more obscure species, and references to historical and other evidence of the former existence of the extinct forms; while Mr. Clarke would like to receive additional information on the birds, with evidence as to the existence of the Great Bustard and other species which once bred in the county, and specimens and information of the occurrence of marine fishes.

LIST OF SUBSCRIBERS.

SIR JOHN LUBBOCK, *Bart.*, *M.P.*, *President of the British Association.*

Duke of Devonshire, *K.G.*, *F.R.S.*,
  Bolton Abbey (two copies).
Earl Fitzwilliam, *K.G.*,
  Wentworth Woodhouse (two copies).
Earl of Wharncliffe, *Wortley Hall*.
Viscount Halifax, *G.C.B.*, *Hickleton*.
Lord Clifton, *M.B.O.U.*,
  Cobham Hall, Gravesend.
Lord Lilford, *Pres. B.O.U.*,
  Lilford Hall, Northants.
Lord Walsingham, *M.A.*,
  Merton Hall, Thetford.

Rev. Thos. Adams, *M.A.*, *York.*
C. M. Adamson, *Newcastle-on-Tyne.*
S. A. Adamson, *F.G.S.*, *Leeds.*
Wm. Aldam, *J.P.*, *D.L.*,
  Frickley Hall.
Tempest Anderson, *M.D.*, *B.Sc.*,
  York.
J. J. Armistead, *Dalbeattie, N.B.*
Edward Atkinson, *F.L.S.*, *Leeds.*
William Atkinson, *Leeds.*

James Backhouse, jun., *York.*
Matthew Bailey, *Flamborough.*
S. D. Bairstow, *F.L.S.*,
  Port Elizabeth, Cape Colony.
Rev. H. F. Barnes-Lawrence, *M.A.*,
  C.M.Z.S., *Birkin, Ferrybridge.*
W. I. Barratt, *Coniston, Lancashire.*
W. Ingram Baynes, *Ulleskelf.*
W. E. Beckwith, *Iron Bridge, Salop.*
James Bedford, *Leeds.*
John T. Beer, *Leeds.*

Thos. W. Belcher, *Whitby.*
Thos. Benn, *Leeds.*
W. S. Bingley, *Leeds* (two copies).
Thos. Birks, jun., *Goole.*
C. Blenkhorn, *Knaresborough.*
Fred. Bond, *F.Z.S.*, *Staines, Middlesex.*
Thos. Boynton, *Ulrome Grange.*
Arthur Briggs, *Rawden, Leeds.*
James Brigham, *Slingsby.*
James Brodie, *Leeds.*
Charles Bromley, *Goole.*
Geo. Brook, ter., *F.L.S.*, *Huddersfield.*
C. J. E. Broughton, *Wortley, Sheffield.*
George Brunton, *Leeds.*
T. E. Buckley, *B.A.*, *F.Z.S.*, *M.B.O.U.*,
  Strathcarron, N.B.
Walter Buckton, *Leeds.*
Thomas Bunker, *Goole.*
J. Darker Butterell, *Beverley.*
J. A. Butterfield, *Wilsden.*

C. A. Carroll, *Boston Spa.*
Godfrey Carter. *Leeds.*
James Carter, *Masham.*
John Henry Carter, *Easington.*
John Henry Carter, jun., *Leeds.*
John Wm. Carter, *Bradford.*
Thos. S. Carter, *Ilkley.*
William Cash, *F.G.S.*, *Halifax.*
Wm. Cheetham, *Horsforth.*
T. R. Clapham, *Austwick Hall.*
J. E. Clark, *B.A.*, *B.Sc.*, *F.G.S.*, *York.*
F. E. Clarke, *Leeds.*
William Clarke, *Leeds.*
T. F. Clayton, *Harrogate.*

Frederick Coates, *Farnley, Leeds.*
G. J. Cockburn, *Headingley, Leeds.*
George Colby, *Malton.*
R. B. Cook. *York.*
James Cooper, *Leeds.*
John Cordeaux, *M.B.O.U.,*
       Great Cotes, Lincolnshire.
R. B. Cragg, *Skipton.*
Henry Crossley, *Wetherby.*
G. H. Crowther, *Wakefield.*

R. D. Darbishire, *B.A., F.G.S.,*
         Manchester.
William Davidson, *Plymouth.*
J. W. Davis, *F.S.A., Halifax.*
Thomas Dawson, *Leeds.*
Francis Day, *F.L.S., Cheltenham.*
Alfred Denny, *Leeds.*
Charles Dixon. *London.*
H. E. Dresser, *F.L.S., M.B.O.U.,*
         London.
W. S. M. D'Urban, *F.L.S., Exeter.*

J. Percy Eddison, *Nottingham.*
J. Ray Eddy, *F.G.S., Carlton, Skipton.*
H. W. T. Ellis, *Crowle, Lincolnshire.*
Thos. W. Embleton, *Methley.*
John Emmet. *Boston Spa.*
W. Hill Evans, *M.D., Bradford.*

Thos. Fairley, *F.R.S.E., Leeds.*
M. T. Farrer, *Ingleborough.*
I G. Featherstone, *Stamford Bridge.*
John Firth, *Bradford.*
William Foggitt, *Thirsk.*
W. A Forbes, *B.A., F.L.S., M.B.O.U.,*
         London.
J. Rawlinson Ford, *Leeds* (two copies).
John Foster, *Lawkland. Clapham.*
Rev. Wm. Fowler, *M.A., Liversedge.*

W. Hodgson Gill, *Hunslet, Leeds.*
Thos. Gough, *B.Sc., F.C.S., York.*
John Grange, *Harrogate.*

John Grassham, *Leeds.*
Rev. Y. Lloyd Greame, *J.P.,*
         Sewerby Hall.
William Gregson, *Baldersby, Thirsk.*
A. C. L. G. Günther, *Ph.D., F.R.S.,*
       British Museum, London.
J. H. Gurney, jun., *F.Z.S., Northrepps.*

Frederick Haigh, *Leeds.*
George Hainsworth, *Leeds.*
John Hancock, *Weybridge, Surrey.*
John Harrison, *Wilstrop Hall.*
W. H. Hay. *Leeds.*
Rhodes Hebblethwaite. *Leeds*
H. Bendelack Hewetson, *Leeds.*
William Hewett, *York.*
Rev. W. C. Hey. *M.A., York.*
J. E. Hill, *London.*
J. Audus Hirst, *Adel, Leeds.*
C. P. Hobkirk, *F.L.S., Huddersfield.*
Benj. Holgate, *F.G.S., Hunslet, Leeds.*
T. Wainman Holmes, *Baildon.*
John Hopkinson, *F.L.S.,*
         Watford, Herts.
H. Knight Horsfield, *Leeds.*
C. Houfton, *Garforth.*
Joseph S. Hurst, *Copt Hewick Hall.*

James Ingleby, *Eavestone, Ripon.*

John Jones, *Hull.*

Arthur R. Kell, *Barnsley.*
Henry Kerr, *Stacksteads, Manchester.*
S. H. Kerr, *Ph.D., M.A., Otley.*
John King, *Leeds.*
Charles Kirkby, *Headingley, Leeds.*

Philip W. Lawton, *Easington, Hull.*
F. Arnold Lees, *F.L.S., Wetherby.*
Arthur R. Linsley. *Leeds.*
C. E. Lister. *Shibden Hall, Halifax.*
William Lobley, *Leeds.*
J. Lucas, *F.G.S., Tooting, Graveney.*

Henry Lupton, *Chapel Allerton, Leeds.*

John McLandsborough, *F.R.A.S.,*
Bradford.

J. C. Mansell-Pleydell, *F.L.S.,*
Blandford, Dorsetshire.

John J. Marsh. *Leeds.*

John Marshall, *Sowerby Bridge.*

Robert Mason, *Glasgow.*

H. T. Mennell, *F.L.S., Croydon.*

R. Morton Middleton, jun., *F.L.S.,*
Castle Eden, co. Durham.

R. Milne-Redhead. *F.L.S.,*
Bolton by Bolland.

F. S. Mitchell, *M.B.O.U., Clitheroe.*

E. A. Moore, *New Wortley, Leeds.*

A. G. More, *F.L.S., M.B.O.U.,*
Dublin.

J. W. Morkill, *Killingbeck, Leeds.*

Walter Morrison, *J.P., Malham Tarn.*

A. S. Myrtle, *M.D., Harrogate.*

W. Naylor, *Whalley, Lancashire.*

T. H. Nelson, *Bishop Auckland.*

F. Nicholson, *F.Z.S., M.B.O.U.,*
Altrincham.

John Norcliffe, *Heckmondwike.*

George Page, *Guisborough.*

G. H. Parke, *F.L.S.,*
Furness Abbey, Lancashire.

I. Patchett, *B.Sc., F.C.S., Birstal.*

R. Lloyd Patterson, *Belfast.*

Jonathan Peel, *Knowlmere, Clitheroe.*

Lister Petty, *Leeds.*

Geo. T. Porritt, *F.L.S., Huddersfield.*

Thomas Pratt, *Ripon.*

Edward E. Prince, *Leeds.*

George A. Prince, jun., *Leeds.*

Mrs. Annie C. Pulleine, *Clifton Castle.*

Frank Ramsden, *Hexthorpe, Doncaster.*

Rev. A. Rawson, *Bromley, Kent.*

F. G. S. Rawson, *Thorpe, Halifax.*

Mrs. Rawson, *Halifax.*

Wm. Eagle Reeves, *London.*

E. G. Reuss. *Sheffield.*

Richard Reynolds, *F.C.S., Leeds.*

Richard Richardson, *Beverley.*

Walter W. Richardson, *Leeds.*

Mrs. Richardson, *Headingley, Leeds.*

Richard Ridehalgh,
Rippenden, Halifax.

Alfred Roberts, *Scarborough.*

George Roberts, *Lofthouse. Wakefield.*

Richard Roberts, *Richmond.*

D. I. Roebuck, *Leeds.*

Walter Rowley, *F.G.S., Leeds.*

James H. Rowntree, *Scarborough.*

John H. Rowntree,
Lord Mayor of York.

W. B. Russell, *LL.B., Leeds.*

John H. Salter, *Lisburn, Ireland.*

Thomas Scattergood, *Leeds.*

Robert Scharff, *Edinburgh.*

Joseph Scott, *Leeds.*

T. Kershaw Skipwith, *Leeds.*

Rev. H. H. Slater, *B.A., F.Z.S.,*
Sharow, Ripon (two copies).

Rev. Henry Smith, *Redcar.*

William Smith, *Morley.*

H. T. Soppitt, *Bradford.*

H. Clifton Sorby, *LL.D., F.R.S.,*
Sheffield.

Thomas Southwell, *F.Z.S., Norwich.*

J. W. Speck, *Headingley, Leeds.*

John Stephenson, *Beverley.*

Thomas Stephenson, *Whitby.*

John Sykes, *M.D., Doncaster.*

Walter Stamper, *Middlesborough.*

Johnson C. Swailes, *Beverley.*

John Tennant, *Leeds.*

Thos. W. Tew, *J.P.,*
Carlton, Pontefract.

J. Campbell Thompson, *Hull.*

Thomas Thorpe, *Pateley Bridge.*

John Thrippleton, *Burley, Leeds.*
George Tindall, *Doncaster.*
Edward Tindall,
    Knapton Hall (two copies).
Thos. Turnbull, *J.P., Whitby.*
W. Barwell Turner, *F.C.S., Leeds.*
Edmund H. Turton,
    Upsall Castle, Thirsk.

Miss Harriet Upton, *Leeds.*

James Varley, *Huddersfield.*

C. Staniland Wake, *Hull.*
G. G. Walmsley, *Liverpool.*
George Ward, *F.I.C., F.C.S., Leeds.*
J. Welburn, *Driffield.*
F. W. T. Vernon Wentworth,
    Wentworth Castle, Barnsley.
William West, *Bradford.*
John Whitaker, *F.Z.S.,*
    Rainworth, Notts.
C. T. Whitmell, *M.A.,B.Sc., Sheffield.*
Joseph Wilcock, *Wakefield.*
John H. Wilson, *Whitby.*
T. Winson, *Spurn Point.*
A. J. H. Wood, *Galphay, Ripon.*

John Wood, *B.A., Boston Spa.*
Basil T. Woodd, *J.P., Knaresborough.*
Fairfax Wooler, *Farnley, Leeds.*
C. A. Wright, *F.L.S., M.B.O.U., Kew.*

British Association (two copies, per the
    local secretaries, York, 1881).
Barnsley Naturalists' Society.
Bradford Naturalists' Society.
Elland-cum-Greetland Naturalists'
    Society.
Giggleswick School Library.
Heckmondwike Naturalists' Society.
The Leeds Library.
Leeds Mechanics' Institution.
Leeds Naturalists' Club.
Leeds Philosophical and Literary
    Society.
Leeds Public Library (two copies).
Liversedge Naturalists' Society.
Chetham Library. Manchester.
Manchester Free Library.
Radcliffe Library. University Museum,
    Oxford.
Scarborough Philosophical Society.
The Yorkshire College, Leeds.
Yorkshire Naturalists' Union.

INDEX

TO THE PRINCIPAL ENGLISH NAMES.